Lecture Notes in Mathematics

A collection of informal reports and seminars
Edited by A. Dold, Heidelberg and B. Eckmann, Zürich

Series: Forschungsinstitut für Mathematik, ETH Zürich

236

Michael Barr
Pierre A. Grillet
Donovan H. van Osdol

Exact Categories and Categories of Sheaves

Springer-Verlag
Berlin · Heidelberg · New York 1971

Michael Barr
University of Fribourg, Fribourg/Switzerland and
McGill University, Montreal/Canada

Pierre A. Grillet
Kansas State University, Manhattan, KS/USA

Donovan H. van Osdol
University of New Hampshire, Durham, NH/USA

AMS Subject Classifications (1970): Primary: 18 B 15, 18 D 99, 18 F 20
Secondary: 18 A 25, 18 C 10

ISBN 3-540-05678-5 Springer-Verlag Berlin · Heidelberg · New York
ISBN 0-387-05678-5 Springer-Verlag New York · Heidelberg · Berlin

Offsetdruck: Julius Beltz, Hemsbach/Bergstr.

Preface

During the summer of 1970, after I had begun the work described in my contribution to this volume, I discovered a preliminary version of what was to become Grillet's. In it he gradually introduced, one at a time, each of the essential hypotheses of exactness (along with some infinite exactness conditions). Although there was little overlap in our results, they seemed to develop a common theme, namely exactness. The strong connection between these papers and later that of Van Osdol suggested making one volume built around the notions of exactness and of sheaves.

At the risk of causing the reader to wonder what I myself have contributed to this volume, I will list the seven mathematicians, three organizations and one typist to whom I am indebted. First there is Saul Lubkin (whom I have never met) but without whose construction I would no doubt still be trying to bound essential monomorphic sequences. Ernie Manes conveyed news to me of what Myles Tierney was doing, in particular the result of Chapter I, section 3, and in addition put me on to the idea of proving an embedding theorem (although he himself later lost interest in the project). Tierney began the idea of studying exact categories. (He soon abandoned it for the greener pastures of elementary toposes.)

Jim Lambek and Jon Beck were frequent listeners and sources of suggestions, some of which were even useful. Heinrich Kleisli suggested the fruitful and important step of going from embedding into M-sets (M a monoid) to embedding into functor categories. Finally, Stephen Chase clarified for me the nature of singular extensions (a notion due originally to Beck). Many of the theorems of Chapter IV are simply copied from his article on Galois objects (although the proofs here are much easier).

The first of three organizations is the National Research Council of Canada (grant NRC - 5281). The second is the Forschungsinstitut für Mathematik of Eidgenössische Technische Hochschule in Zürich which provided both a living during the summer of 1970 and the platform from which almost all of this theory was exposed during the year 1970 - 1971. The Fonds National Suisse (project no. 2.180.69) provided support during the accademic year 1970 - 1971 through a grant to the

University of Fribourg.

Finally, I would like to thank Frau Hildegard Mourad for her skill in deciphering my handwriting and the patience and care with which she typed my manuscript. Van Osdol would like to make a similar acknowledgement to Miss Jean Gahan.

Michael Barr

Fribourg, June 1971

Table of Contents

Chapter V. Cohomology

Pierre Antoine Grillet: Regular Categories

Donovan H. van Osdol: Sheaves in Regular Categories

EXACT CATEGORIES

by Michael Barr

Introduction

Exact categories, roughly speaking, are categories which satisfy the equation

$$(\text{Abelian}) = (\text{Exact}) + (\text{Additive}).$$

Generally speaking, the axioms of abelian categories were chosen precisely in order to define a good notion of the homology theory of chain complexes of a category If one wishes to remove additivity, there are two possible directions. One direction is to try to axiomatize non-abelian homology. This leads to consideration of pointed categories and then of normal monomorphisms and epimorphisms — those which are kernels and cokernels, respectively. This is essentially the point of view adopted by Brinkmann and Puppe in [BP] and Gerstenhaber-Moore in [Ge]. In essence, it goes back at least as far as Mitchell ([Mi], I.15). Brinkmann and Puppe even use the term exact category to describe the type of categories they are considering. Gerstenhaber does not name the type of categories he is dealing with. His axioms are related to but somewhat different from those of Brinkmann and Puppe. Both suppose as part of their axioms that normal epimorphisms are invariant under pullback. I do not know a single example of a category satisfying that hypothesis unless it also satisfies the hypothesis that every regular epimorphism is normal. A regular epimorphism is one which is the coequalizer of some pair of maps and it is evident that every normal epimorphism is regular, since it is the coequalizer of 0 and whatever it is the kernel of. But the nicest pointed category of all, pointed sets, does not satisfy this assumption, in sharp contrast of the result of Manes [Man], that every additive equational category is abelian. In addition, I have been unable to decide, after a modest expenditure of time, whether the categories of monoids and commutative monoids satisfy the Gerstenhaber-Moore axioms. This is one motivation for ignoring earlier definitions of exactness. A second is the essentially special nature of non-abelian cohomology. Its interest is practically restricted to categories which are more or less like groups. I feel that the term exact is too basic to be used for such a special theory.

The second approach is in the direction of homotopy. By the theorem of Dold-Puppe ([DP], Chapter 3), in an abelian category chain complexes (concentrated in non-negative degrees) are equivalent to simplicial objects. This suggests, at least, that one fruitful direction of inquiry is to find a good theory of homotopy for simplicial objects. It would also be nice if every equational category satisfied the conditions and, of course, if it satisfied the above equation.

The exact categories defined here have precisely these properties. It all began with a theorem of Tierney (unpublished, but see I.(3.11)[1] below) that a category is abelian if and only if and only if it is additive and has finite limits and colimits and universally effective equivalence relations. The definition of exact category given here is a slight weakening of the above, weakened only for technical reasons. An exact category has certain finite limits and colimits and universally effective equivalence relations (see I. (1.2) and I. (1.3) for definitions).

The contents of this paper include the elementary properties of

---

[1] A reference of the form N. (a.b) is to Chapter N̲ , paragraph (a,b). A reference of the form (a.b) is the same chapter, paragraph (a.b).

exact categories (I and II), an embedding and meta-theorem which generalize those of Mitchell ([Mi] VI, theorem 1.2) in the abelian case (III), and an application to cohomology and Baer addition of extensions (IV and V). The simplicity of the presentation of the Baer sum should be compared with that of Gerstenhaber in [Ge]. The completeness of the results should be compared with those of Chase in [Ch] in which an unpleasant and unnatural assumption ("coflatness") had to be introduced for want of the notion of right exact sequences.

The homotopy theory is not at all developed here. It is possible, given a simplicial object in an exact category, to say when that is a Kan object; and when it is, to define its homotopy. This will be the subject of a subsequent work. The homotopy so defined will be an object of the category in question, rather than a group. It is base-point free and in sets is the usual groupoid (except in dimension 0) of homotopy classes of maps of spheres. The usual homotopy is recovered as soon as a principal component and a base point there are chosen.

There is one more point I would like to mention. A useful axiom which gives a notion intermediate between being exact and being abelian is the supposition that every reflexive subobject of the square of any object is an equivalence relation (see I. (5.5)). This condition is equivalent to every simplicial object being Kan. It is also sufficient to have the theory of group actions of Chapter IV work equally well for monoid actions. The theory of monoid actions also works well in the category of sets, but for an entirely different reason: that category is cartesian closed so that cartesian products commute with all colimits.

Chapter I. The Elementary Theory

1. Definitions and examples.

(1.1) One of the most important tools will be the factorization of every morphism as a regular epimorphism followed by a monomorphism (see (2.3) below). A regular epimorphism is a map which is the co-equalizer of some pair of maps, which can be supposed to be its kernel pair, if that exists. We adopt (or adapt) the notation of MacLane [Mac] and we use $\rightarrowtail$ to denote a monomorphism, $\twoheadrightarrow$ to denote a regular epimorphism, and $\xrightarrow{\sim}$ to denote an isomorphism. We will also use these arrows as substantives and say, for example, "f is $\rightarrowtail$" to mean that f is a monomorphism.

(1.2) If $f: X \to X'$ is any map in any category, its kernel pair $X'' \rightrightarrows X$ has the property that $(-,X'') \rightarrowtail (-,X) \times (-,X)$ is a natural equivalence relation on $(-,X)$; two maps to X are identified if and only if their compositions with f are equal. In general, two maps $X'' \rightrightarrows X$ for which $(-,X'') \rightarrowtail (-,X) \times (-,X)$ is a natural equivalence relation on $(-,X)$ will be called on equivalence relation on X. Not every equivalence relation on X need be a kernel pair, any completeness hypothesis notwithstanding. See (1.4) example 5 below. An equivalence relation which is a kernel pair will be called effective.

(1.3) Let $\underline{X}$ be a category. We say that $\underline{X}$ is <u>regular</u> if it satisfies EX1) below and <u>exact</u> if it satisfies EX2) in addition.

(EX1) The kernel pair of every map exist and have a coequalizer; moreover every diagram of the form

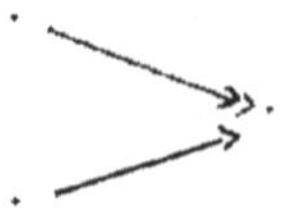

has a coequalizer which is of the form

EX2) Every equivalence relation is effective.

(1.4) The following are examples of regular categories. All are exact except example 5.

1. The category $\underline{S}$ of sets.
2. The category of non-empty sets.
3. For any triple $\mathbb{T}$ on $\underline{S}$, the category $\underline{S}^{\mathbb{T}}$ of $\mathbb{T}$-algebras.
4. Every partially ordered set considered as a category.
5. The category of Stone spaces (compact hausdorff 0-dimensional spaces).
6. Any abelian category.
7. For any small category $\underline{C}$, the functor category $(\underline{C}^{op},\underline{S})$.
8. For any topology on $\underline{C}$, the category $\mathcal{S}(\underline{C}^{op},\underline{S})$ of sheaves.

(1.5) Remark. It should be noted that unlike the notion of abelianness, exactness is not self-dual. Outside of abelian categories and the categories of sets and pointed sets, the only category that I know of which is tripleable over $\underline{S}$ and both exact and coexact is compact hausdorff spaces (and its dual, C*-algebras).

(1.6) Definition. Let $\underline{X}$ be a regular category. A sequence

$$X' \overset{d^o}{\underset{d^1}{\rightrightarrows}} X \xrightarrow{d} X''$$

is called

a) left exact if (d^o,d^1) is the kernel pair of d;

b) right exact if d is the coequalizer of d^o and d^1, and, moreover the image of (d^o, d^1) in $X \times X$ is the kernel pair of d (see (2.1) and (2.4) below);

c) exact if it is both left and right exact.

(1.7) Definition. Let $\underline{X}$ and $\underline{Y}$ be exact categories. A functor U: $\underline{X} \to \underline{Y}$ is called

a) quasi-exact it it preserves exact sequences;

b) exact if, in addition, it preserves all finite limits;

c) reflexively (quasi) exact if it is (quasi) exact and reflects isomorphisms.

(1.8) Examples. The following are examples of exact functors.

1. For any triple on $\underline{S}$, the underlying functor $\underline{S}^{\mathbb{T}} \longrightarrow \underline{S}$.
2. For any small category $\underline{C}$ and any object of $\underline{C}$, the functor $(\underline{C}^{op}, \underline{S}) \to \underline{S}$ which evaluates a functor at C. Of course this functor preserves all limits and colimits.
3. For any topology on $\underline{C}$, the associated-sheaf functor $(\underline{C}^{op}, \underline{S}) \to \mathfrak{F}(\underline{C}^{op}, \underline{S})$.
4. Any (additive) exact functor between abelian categories.

Of these examples, only 1 is reflexively exact in general.

2. Preliminary results.

(2.1) Throughout this section, $\underline{X}$ denotes a regular category. We will establish some of its basic properties, in particular the factorization.

(2.2) Proposition. Suppose $X \twoheadrightarrow Y \to Z$ is given. Then $X \times_Z X \to Y \times_Z Y$ is an epimorphism.

Proof. The diagrams

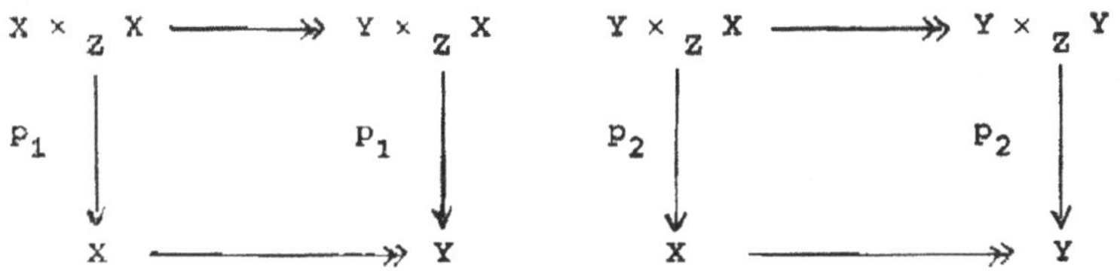

are each easily seen to be pullbacks, where p_1 and p_2 are the respective coordinate projections. A composite of two $\twoheadrightarrow$ is certainly an epimorphism and, as we will see in (2.8), is $\twoheadrightarrow$.

(2.3) Theorem. Every map has a factorization of the form $\cdot \twoheadrightarrow \cdot \rightarrowtail \cdot$.

Proof. Begin with a map $X \to Z$, form its kernel pair, and let Y be their coequalizer. There is induced a map $Y \to Z$ and we can form its kernel pair to get

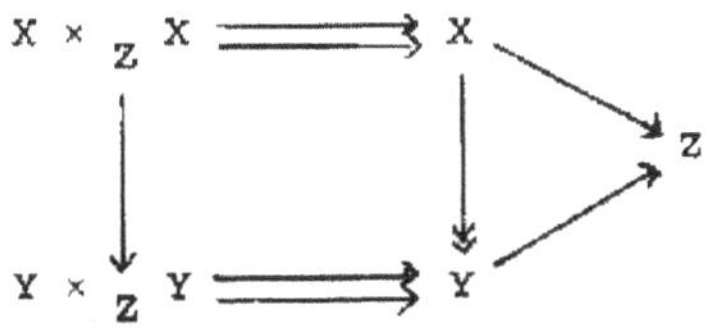

From the fact that $X \to Y$ coequalizes $X \times_Z X \rightrightarrows X$ and that $X \times_Z X \to Y \times_Z Y$ is an epimorphism, it follows that the two projections $Y \times_Z Y \rightrightarrows Y$ are equal and that $Y \rightarrowtail Z$. Thus the map is

factored

$$X \twoheadrightarrow Y \rightarrowtail Z.$$

(2.4) Remark. With minor modifications, this is essentially a theorem of Kelly's ([Ke], proposition 4.2). It is clear that to prove it one need only suppose that a pullback of a regular epimorphism is an epimorphism.

(2.5) Proposition. If the composite f.g is $\twoheadrightarrow$, so is f.

Proof. If f.g is the coequalizer of d^0 and d^1, than f is the coequalizer of $g.d^0$ and $g.d^1$.

(2.6) Proposition. Every commutative diagram

has a diagonal map as indicated so that both triangles commute

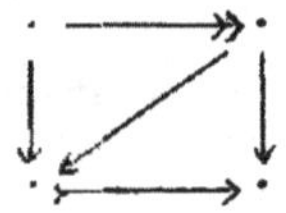

Proof. Consider the diagram

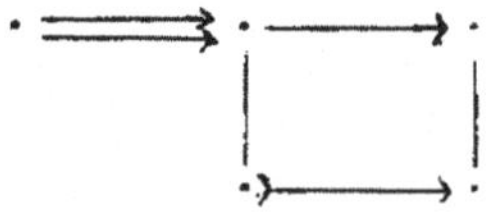

in which the top row is a coequalizer.

(2.7) Corollary. Any map which is both $\rightarrowtail$ and $\twoheadrightarrow$ is $\xrightarrow{\sim}$.

Proof. Consider

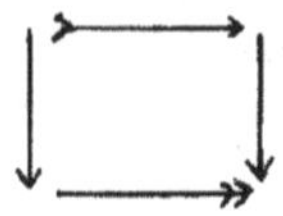

where the top and bottom are the given map and the vertical maps are identities.

(2.8) <u>Corollary</u>. If $\cdot \xrightarrow{f}\!\!\!\!\twoheadrightarrow \cdot \xrightarrow{g}\!\!\!\!\twoheadrightarrow \cdot$, then $\xrightarrow{gf}\!\!\!\!\twoheadrightarrow \cdot$

Proof. Factor gf as $\cdot \xrightarrow{h}\!\!\!\!\twoheadrightarrow \cdot \overset{k}{\rightarrowtail} \cdot$ and consider

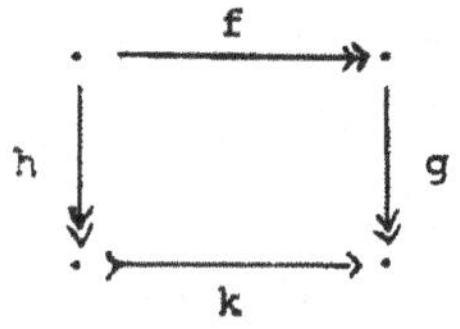

The existence of a diagonal presents k as the second factor of a $\twoheadrightarrow$, whence k is $\twoheadrightarrow$ also, by (2.5), and hence an $\xrightarrow{\sim}$.

(2.9) <u>Corollary</u>. The factorization of (2.3) is unique up to a unique $\xrightarrow{\sim}$.

Proof. Two applications of (2.6).

(2.10) <u>Proposition</u>. An exact functor preserves factorizations.

Proof. A right exact functor evidently preserves $\twoheadrightarrow$ and a left exact functor, by preserving the pullback of $\rightarrowtail$ (which has a limit = dom(f) if and only if f is $\rightarrowtail$), preserves $\rightarrowtail$. Thus it takes the $\cdot \twoheadrightarrow \cdot \rightarrowtail \cdot$ factorization into one which by uniqueness is the required factorization.

(2.11) <u>Proposition</u>. Let $\underline{X}$ and $\underline{Y}$ be exact, $X'' \rightrightarrows X \longrightarrow X'$ a left (resp. right) exact sequence, and U an exact functor. Then $UX'' \rightrightarrows UX \longrightarrow UX'$ is left (resp. right) exact.

Proof. The left half of this is pretty clear. As for the right, let $X_0 \rightarrowtail X \times X$ be the image of $X'' \longrightarrow X \times X$. Then we have

$$X'' \twoheadrightarrow X_0;\ X_0 \rightrightarrows X \twoheadrightarrow X'$$

in which the second is exact. Applying U we have

$$UX'' \twoheadrightarrow UX_0;\quad UX_0 \rightrightarrows UX \longrightarrow UX'$$

in which the second is exact. But this readily implies that

$$UX'' \rightrightarrows UX \longrightarrow UX'$$

is right exact.

(2.12) <u>Remark</u>. It was to make true this proposition (whose proof is the same as of II, proposition 4.3 of [CE]) that the somewhat unusual definition of right exact sequence was chosen.

(2.13) <u>Proposition</u>. In order that $X' \rightrightarrows X \longrightarrow X''$ be exact, it is necessary and sufficient that $X \xrightarrow{f} X''$ (↠) and $X' \rightrightarrows X$ be its kernel pair.

Proof. It is clearly necessary. But if f is $\twoheadrightarrow$, then it is evidently the coequalizer of its kernel pair.

(2.14) <u>Corollary</u>. A functor is quasi-exact if it preserves kernel pairs and $\twoheadrightarrow$; it is exact if it preserves all finite limits and $\twoheadrightarrow$.

(2.15) <u>Proposition</u>. If the product of a finite number of exact sequences exists, it is exact.

Proof. Since a product of kernel pairs is a kernel pair, it is sufficient to show that a product of $\twoheadrightarrow$ is again $\twoheadrightarrow$. Suppose $X \twoheadrightarrow X'$ and $Y \twoheadrightarrow Y'$. As soon as $X' \times Y'$ exists, so do $X \times Y'$ and $X \times Y$, since each of the squares below is a pullback. The vertical arrows are the evident coordinate projections,

$$\begin{array}{ccc} X \times Y & \twoheadrightarrow & X \times Y' \\ \downarrow & & \downarrow \\ Y & \twoheadrightarrow & Y' \end{array} \qquad \begin{array}{ccc} X \times Y' & \twoheadrightarrow & X' \times Y' \\ \downarrow & & \downarrow \\ X & \twoheadrightarrow & X' . \end{array}$$

Composing, we have $X \times Y \twoheadrightarrow X' \times Y'$.

(2.16) Corollary. For any object X of the exact category $\underline{X}$, $X\times -: \underline{X} \longrightarrow \underline{X}$ is a quasi-exact functor (provided it exists).

Proof: $X \rightrightarrows X \longrightarrow X$ (all maps being identity) is exact.

(2.17) Corollary. Let $\underline{X}$ have finite powers. For any finite integer n, the cartesian n-th power functor $\underline{X} \longrightarrow \underline{X}$ is exact.

Proof. Clear from (2.15) and the fact limits commute with each other.

(2.18) Remark. If the cartesian n-~~th~~ power functor exists and preserves $\twoheadrightarrow$ for all cardinals n or for all $n < N_0$, then that functor is exact for all such n.

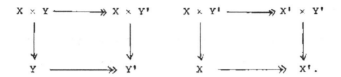

3. Additive exact categories.

(3.1) This section is devoted to proving Tierney's theorem that a non-empty additive exact category is abelian. Throughout this section $\underline{A}$ denotes such a category; $\underline{Ab}$ denotes the category of abelian groups.

(3.2) Let $A \in \underline{A}$, and consider any 0 map, say $0: A \longrightarrow A$. Since 0 coequalizes any two maps, the kernel pair of this is $A \times A$, which then exists. Let Z be the coequalizer of the projections

$$A \times A \rightrightarrows A \longrightarrow Z.$$

For any $B \in \underline{B}$,

$$(Z,B) \longrightarrow (A,B) \rightrightarrows (A \times A,B) \simeq (A,B)\times(A,B)$$

is an equalizer, which implies, since all these homs take values in $\underline{Ab}$, $(Z,B) = 0$. In an additive category, any initial object is a zero object, and so $Z = 0$. Moreover, A was an arbitrary object and we showed that $A \twoheadrightarrow 0$. Thus we have proved

(3.3) Proposition. There is a zero object 0 and $A \twoheadrightarrow 0$ for any A.

(3.4) Corollary. Finite products exist in $\underline{A}$.

Proof. For any $A,B \in \underline{A}$,

$$\begin{array}{ccc} A \times B & \longrightarrow & A \\ \downarrow & & \downarrow \\ B & \twoheadrightarrow & 0 \end{array}$$

is a pullback.

(3.5) Proposition. Maps in $\underline{A}$ have kernels.

Proof. Let $f: A \longrightarrow A'$. From the kernel pair $A'' \underset{d^1}{\overset{d^0}{\rightrightarrows}} A$ and let $s: A \longrightarrow A''$ be the diagonal map. I claim that $A'' \xrightarrow{d^0-d^1} A$ is a weak kernel. First, $f.(d^0-d^1) = fd^0-fd^1 = 0$. Second, if $g: B \longrightarrow A$ is such that $f.g = 0$, let $k: B \longrightarrow A''$ be such that $d^0.k = g$ and $d^1.k = 0$. Then $(d^0 - d^1).k = g$. It is clear that the image of $d^0 - d^1$ must be the kernel.

(3.6) Corollary. $\underline{A}$ has finite limits.

Proof. It is well-known that in an additive category kernels and finite products are enough.

(3.7) Proposition. Let A be an object of $\underline{A}$ and $A' \rightarrowtail A \times A$, containing the diagonal of A. Then A' is an equivalence relation on A.

Proof. The property of being an equivalence relation is defined with respect to the representable functors, which can be considered to take values in $\underline{Ab}$. But then $(-,A') \rightarrowtail (-,A)\times(-,A)$ will still contain the diagonal. In $\underline{Ab}$ the assertion is trivial and the above argument shows it is true for any additive category.

(3.8) Proposition. Every monomorphism of $\underline{A}$ is normal (that is, a kernel).

Proof. Let $A' \overset{f}{\rightarrowtail} A$. Form

$$A' \times A \underset{\binom{0}{1}}{\overset{\binom{f}{1}}{\rightrightarrows}} A.$$

It is easily seen that the induced map $\begin{pmatrix} f & 0 \\ 1 & 1 \end{pmatrix} : A' \times A \rightrightarrows A \times A$ is $\rightarrowtail$ and contains the diagonal, and hence is an equivalence relation and therefore a kernel pair. But it is clear that a map coequalizes $\binom{f}{1}$ and $\binom{0}{1}$ if and only if it annihilates f so that that coequalizer of those maps is the cokernel of f. Conversely, $\binom{f}{1}$ and $\binom{0}{1}$ being the kernel pair of that cokernel is equivalent to f being its kernel.

Notice that in the course of this proof we have shown that every

$\rightarrowtail$ has a cokernel, which implies, by the standard factorization, that every map does. The finite products are also coproducts. An additive category is cocomplete as soon as it has direct sums and coequalizers. Thus we have:

(3.9) Proposition. $\underline{A}$ is finitely cocomplete.

(3.10) Proposition. Every epimorphism in $\underline{A}$ is normal.

Proof. Let f be an epimorphism and factor it as $\cdot \xrightarrow{g} \!\!\!\!\rightarrow \cdot \xrightarrow{h} \cdot$ (with h a monomorphism $\rightarrowtail$). Since h is normal, it is the kernel of some k. If $k \neq 0$, we would have $kf = 0$, which contradicts f being an epimorphism. Thus h is an isomorphism, which means that f is $\twoheadrightarrow$. In an additive category this implies that f is normal.

(3.11) Theorem. (Tierney). $\underline{A}$ is abelian.

Proof. $\underline{A}$ is additive; it is finitely complete and cocomplete; every map has a factorization as an epimorphism followed by a monomorphism; every monomorphism and every epimorphism is normal.

(3.12) Example. The category of torsion free abelian groups is regular, but not exact.

4. Regular epimorphism sheaves.

(4.1) If $\underline{C}$ is a category, a collection of families $\{U_i \longrightarrow U \mid i \in I\}$ (called coverings) is called a Grothendieck topology on $\underline{C}$ (see [Ar], I, Definition (0.1)), if it satisfies the following conditions.

a) Every $\{U \xrightarrow{f} U'\}$ with f an isomorphism is a covering.

b) If $\{U_i \longrightarrow U \mid i \in I\}$ is a covering and for each $i \in I$, $\{U_{ij} \longrightarrow U_i \mid j \in I_i\}$ is a covering, so is $\{U_{ij} \longrightarrow U \mid i \in I,\ j \in I_i\}$.

c) If $\{U_i \longrightarrow U \mid i \in I\}$ is a covering and $V \longrightarrow U$ is a map, each of pullbacks $U_i \times_U V$ exists and

$$\{U_i \times_U V \longrightarrow V \mid i \in I\}$$

is a covering.

It is easily seen from EX1) and (2.8) that these conditions are satisfied if we take for coverings exactly the $U' \twoheadrightarrow U$. This will be called the regular epimorphism topology. The axiom of a regular category might almost have been chosen with this topology in mind.

(4.2) Given a topology on $\underline{C}$ as above, a sheaf of sets on $\underline{C}$ is a functor $F: \underline{C}^{op} \longrightarrow \underline{S}$ such that for every covering $\{U_i \longrightarrow U \mid i \in I\}$,

$$FU \longrightarrow \prod_{i \in I} FU_i \rightrightarrows \prod_{i,j \in I} F(U_i \times_U U_j)$$

is an equalizer. The category of sheaves (with natural transformations as morphisms) is denoted $\mathfrak{F}(\underline{C}^{op}, \underline{S})$. It is equipped with a full faithful embedding $\mathfrak{F}(\underline{C}^{op}, \underline{S}) \longrightarrow (\underline{C}^{op}, \underline{S})$ which has an exact left adjoint. Conversely any coreflective subcategory $\underline{E}$ of a set-valued functor category $(\underline{C}^{op}, \underline{S})$ with an exact coreflector (left adjoint for inclusion) will be a category $\mathfrak{F}(\underline{D}^{op}, \underline{S})$ for some $\underline{D}$ and some Grothendieck topology on $\underline{D}$ for which each of the representable functors is a sheaf. (Such a topology is said to be less fine than the canonical topology; the

the canonical topology is the finest topology for which all representable functors are sheaves.[1]) Evidently $\underline{D}$ may be taken to be $\underline{C}$ iff each of the representable functors of $(\underline{C}^{op},\underline{S})$ is in $\underline{E}$. Such an $\underline{E}$ is called a topos. [1]

(4.3) Proposition. Let $\underline{X}$ be a small regular category. Let $\mathfrak{F}(\underline{X}^{op},\underline{S})$ denote the category of set valued sheaves for the regular epimorphism topology described above. Then the canonical embedding $\underline{X} \longrightarrow \mathfrak{F}(\underline{X}^{op},\underline{S})$ is full, faithful and exact.

Proof. It is clear that this topology is less fine than the canonical one, so the Yoneda embedding of $\underline{X}$ takes it into sheaves. The embedding preserves all limits, since the Yoneda embedding does, and it is well known that the embedding of sheaves into all functors creates limits. It is full and faithful for the same reason. Finally, a sheaf F, evaluated at an exact sequence

$$X' \times_X X' \rightrightarrows X' \longrightarrow X,$$

must produce an equalizer

$$FX \longrightarrow FX' \rightrightarrows F(X' \times_X X'),$$

according to the definition of sheaf. By the Yoneda lemma, this is

$$((-,X)F) \longrightarrow ((-,X'),F) \rightrightarrows ((-,X' \times_X X'),F)$$

and that sequence being an equalizer is the some as

$$(-,X' \times_X X') \rightrightarrows (-,X') \longrightarrow\!\!\!\!\!\rightarrow (-,X)$$

being a coequalizer in this particular subcategory of the functor category.

(4.5) From this proposition we see that regular categories may be characterized as categories having kernel pairs, pullbacks along regular epimorphisms, coequalizers of kernel pairs, and for every small

1 See Appendix for an improved statement and proof of this result.

full subcategory stable under these operation, a full exact embedding into a topos. The converse is clear. A topos is complete and cocomplete and even exact. If our given category is itself small, we can replace it by its finite limit completion in its embedding into a topos and suppose it has finite limits.

5. Constructions on regular and exact categories.

(5.1) In this section $\underline{X}$ represents a regular (resp. exact) category. We are going to describe two types of constructions which when applied to $\underline{X}$ automatically produce another regular (resp. exact) category.

(5.2) Let $\underline{I}$ be an arbitrary category and D: $\underline{I} \longrightarrow \underline{X}$ a functor. We will say that the pair (D,$\underline{I}$) or D alone is a diagram in $\underline{X}$. Note that $\underline{I}$ is not required even to be small. The comma category ($\underline{X}$,D) has for objects pairs (X,α), where X is an object of $\underline{X}$ and α is a natural transformation from (the constant functor whose value is) X to D. A morphism of ($\underline{X}$,D) is a morphism f in $\underline{X}$ giving a commutative triangle

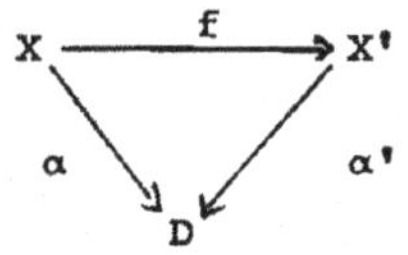

(5.3) Proposition. The forgetful functor $(\underline{X},D) \longrightarrow \underline{X}$, which takes $(X,\alpha) \longmapsto X$, creates whatever colimits exist in X as well as kernel pairs, pullbacks, finite monomorphic families, and the limit of any diagram E: $\underline{J} \longrightarrow \underline{X}$ in which $\underline{J}$ has a terminal object (and in which the limit exists, of course).

Proof. Given a diagram E: $\underline{J} \longrightarrow (\underline{X},D)$ which has a colimit in $\underline{X}$, the universal mapping property of colimit will endow that object with a map to D. As for limits, supposing $\underline{J}$ has a terminal object j_0, a functor E: $\underline{J} \longrightarrow (\underline{X},D)$ is precisely given by a functor E: $\underline{J} \longrightarrow \underline{X}$ together with a natural transformation $Ej_0 \longrightarrow D$. This determines the lifting of E to ($\underline{X}$, D). The limit $X \longrightarrow E$, when it exists, will equally have a unique map $X \longrightarrow Ej_0 \longrightarrow D$ which lifts X into ($\underline{X}$,D). It is now trivial to see that X is the limit there also. If $f_1,\dots,f_n$: $X \longrightarrow Y$ is a finite (or for that matter infinite) set of maps, it is called a monomorphic family if for all Z and maps $g.h\colon Z \longrightarrow X$, $f_i.g = f_i.h$ for $i = 1,\dots,n$ implies that $g = h$. If $Y \longrightarrow D$ is given and $f_1,\dots,f_n\colon X \longrightarrow Y$ are all maps over D, then they are simultaneously coequalized by $Y \longrightarrow D$. If they do not form a monomorphic family in $\underline{X}$, then there are $g \neq h\colon Z \longrightarrow X$ with $f_i.g = f_i.h$ for $i = 1,\dots,n$. Then all the composites $Z \overset{h}{\underset{g}{\rightrightarrows}} X \xrightarrow{f_i} Y \longrightarrow D$ are the same. Thus $g \neq h$ as maps over D, and so $\{f_i\}$ is not a monomorphic family in $(\underline{X},D)$ either.

(5.4) Theorem. Let $\underline{X}$ be regular (resp. exact) and $D\colon \underline{I} \longrightarrow \underline{X}$ a functor. Then $(\underline{X},D)$ is regular (resp. exact).

Proof. Everything except exactness follows from (5.3) and the easily proved (from (5.3)) assertion that $(\underline{X},D) \longrightarrow \underline{X}$ preserves $\longrightarrow\!\!\!\!\!\rightarrow$. Exactness (when $\underline{X}$ is exact) also follows from (5.3) if we can show that the underlying functor preserves equivalence relations. To do this we show the following combinatorial characterization of equivalence relations.

(5.5) Proposition. Let $\underline{X}$ be a category which has pullbacks of split epimorphisms. Then $X \overset{d^o}{\underset{d^1}{\rightrightarrows}} Y$ is an equivalence relation if and only if the following conditions are satisfied.

a) $X \overset{d^o}{\underset{d^1}{\rightrightarrows}} Y$ is a monomorphic family.

b) There is an $r\colon Y \longrightarrow X$ such that $d^o.r = d^1.r = Y (= \text{id } Y)$.

c) There is an $s\colon X \longrightarrow X$ such that $d^o.s = d^1$ and $d^1.s = d^o$.

d) In the diagram below in which Z is a pullback of d^o and d^1, there is a map t as indicated making each of the outside squares commutative.

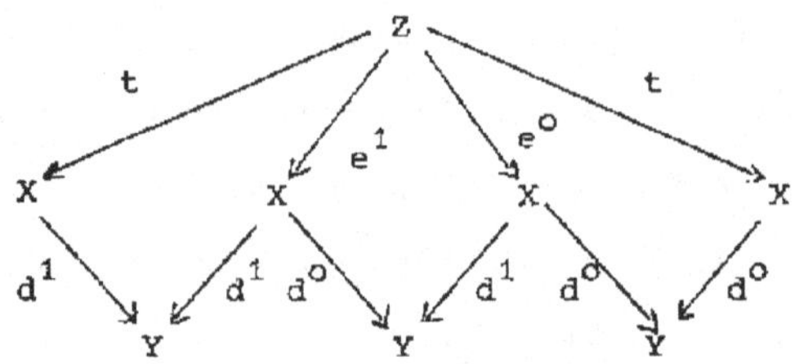

Proof. I leave it as an exercise to show that in $\underline{S}$, the existence of r,s,t translates the usual reflexive, symmetric, and transitive laws and hence the existence of (–,r), (–,s), (–,t) will show that (–,X) is an equivalence relation on (–,Y). To go the other way, suppose $X \underset{d^1}{\overset{d^o}{\rightrightarrows}} Y$ is an equivalence relation. Then $(Y,X) \longrightarrow (Y,Y)\times(Y,Y)$ must contain the diagonal of (Y,Y), so in particular the diagonal element (idY, idY) and the $r \in (Y,X)$ mapping to it is the required map. $(X,X) \longrightarrow (X,Y)\times(X,Y)$ is symmetric, and since (d^o,d^1) is in the image of (X,X) (it is the image of the identity map), so must (d^1,d^o) be. The element of (X,X) having those projections is s . Finally letting Z be the pullback as above, we observe that $(Z,X) \longrightarrow (Z,Y)\times(Z,Y)$ is transitive. In particular the images of e^o and e^1 are $(d^o.e^o,\ d^1.e^o)$ and $(d^o.e^1,\ d^1.e^1)$ respectively, and the equation $d^1.e^o = d^o.e^1$ implies the existence of t with projections $d^o.e^o$ and $d^1.e^1$, exactly as required.

(5.6) Corollary. Suppose $\underline{X}$ has, and a functor $U: \underline{X} \longrightarrow \underline{Y}$ preserves pullbacks along split epimorphisms; in addition suppose U preserves monomorphic pairs of maps. Then U preserves equivalence relations.

Proof. Trivial.

(5.7) Let Th be any finitary algebraic theory. This means Th is a category with a functor $n \longmapsto (n)$ from the category of finite sets which preserves coproduct $((n)+(m) = (n+m))$ and is an isomorphism on objects. The category $\underline{S}^{\underline{Th}}$ is the category of product preserving functors $\underline{Th}^{op} \longrightarrow \underline{S}$. Included are all the familiar categories of algebra — in particular groups and abelian groups. If $\underline{X}$ is an arbitrary category, $\underline{X}^{\underline{Th}}$ can be defined as the category whose objects consists of objects $X \in \underline{X}$ together with a lifting of the hom functor $(-,X)$: $\underline{X}^{op} \longrightarrow \underline{S}$ into $\underline{S}^{\underline{Th}}$. A morphism between two such objects is a natural transformation between these functors. Since $\underline{S}^{\underline{Th}} \longrightarrow \underline{S}$ is faithful, this is equivalent, by the Yoneda lemma, to a map between the objects which induces $\underline{S}^{\underline{Th}}$ morphisms on the hom sets. When $\underline{X}$ itself has finite products, it is well known that an algebra is also equivalent to a product preserving functor $\underline{Th}^{op} \longrightarrow \underline{X}$. Moreover this condition is "local" in the sense that in order to recover the equivalence it is only necessary to know the algebra structure for a few objects, namely the powers of $\underline{X}$. For example, a group structure on X is either given by a lifting of $(-,X)$ through the category of groups or by giving morphisms $1 \longrightarrow X$, $X \longrightarrow X$, $X \times X \longrightarrow X$ satisfying laws of a group unit, inverse, and multiplication, respectively (1 denotes the terminal object or 0th power). These morphisms are found by observing that $(1,X)$, (X,X) and $(X \times X,X)$ have group structures. The unit of the first, the inverse (under the group law!) of the identity of X in the second, and the product of the two projections in the third of these groups are the required mappings. However, as the next proposition and its corollary show, when the theory has nullary operations (e.g. groups), then we may as well suppose it has products and the two descriptions coincide. A nullary operation is a map in $\underline{Th}$ of $1 \longrightarrow 0$ and entails for any an algebra X an "element" of $(-,X)$. This means a natural transformation of the constant functor 1 to $(-,X)$. Equivalently it assigns to each Y an $\alpha Y\colon Y \longrightarrow X$ such that for $f\colon Y \longrightarrow Y'$, $\alpha Y'.f = \alpha Y$.

(5.8) <u>Proposition</u>. Let an object $X \in \underline{X}$ admit a constant operation. Then x has a terminal object.

Proof. Choose Y arbitrarily and factor αY as $Y \xrightarrow{\beta Y} T \rightarrowtail X$. If we also factor αX as $X \twoheadrightarrow T_o \rightarrowtail X$, then the diagonal fill-in of the diagram

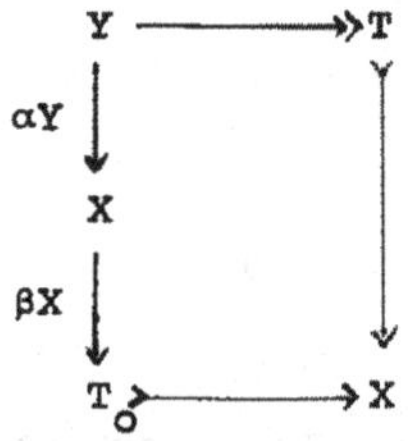

which commutes by naturality of α, gives that $T \rightarrowtail T_o$ and that every object has at least one map to T_o which factors αY. Naturality gives $\alpha T_o . \beta X = \alpha X$. Since we gave αX its unique factorization as βX followed by inclusion of T_o, it follows that αT_o is that inclusion. Finally, for any $f\colon Y \longrightarrow T_o$, $\alpha T_o . f = \alpha Y$, and we may cancel αT_o to conclude that f is $Y \xrightarrow{\beta Y} T \rightarrowtail T_o$, which means that Y has only one map to T_o.

(5.9) <u>Corollary</u>. Every object of $\underline{X}$ has finite powers.

Proof. Once there is a terminal object 1, the kernel pair of $X \longrightarrow 1$ is $X \times X$. Higher products may be constructed by pulling back along coordinate projections

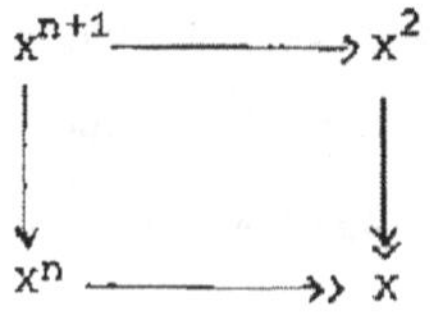

which are $\longrightarrow\!\!\!\!\rightarrow$ (split by the diagonal map).

(5.10) *Proposition*. Let $\mathcal{F}(\underline{X}^{op},\underline{S})$ be the category of set valued sheaves in the regular epimorphism topology (4.1). Let $\underline{Th}$ be a finitary theory. Then the functor $X \longmapsto (-,X)$ preserves $\underline{Th}$ objects and $\underline{Th}$ morphisms.

Proof. The inclusion of sheaves into the whole functor category preserves limits, so the products given in the proof are the products as sheaves. If X is a $\underline{Th}$ object in $\underline{X}$, this means there is, for each $(n) \longrightarrow (m)$ in $\underline{Th}$, a map $(Y,X)^m \longrightarrow (Y,X)^n$ which is natural in Y. Corresponding to each commutative diagram

$$\begin{array}{ccc} & (m) & \\ \swarrow & & \searrow \\ (n) & \longrightarrow & (p) \end{array}$$

the diagram

$$\begin{array}{ccc} (Y,X)^p & \longrightarrow & (Y,X)^n \\ & \searrow \quad \swarrow & \\ & (Y,X)^m & \end{array}$$

must also commute. Everything being natural in Y, this means that there is a natural transformation

$$(-,X)^m \longrightarrow (-,X)^n$$

for each $(n) \longrightarrow (m)$ in $\underline{Th}$ such that diagrams corresponding to the above commute. That is, we have a product preserving functor, $m \longmapsto (-,X)^m$ of $\underline{Th}^{op} \longrightarrow \mathcal{S}(\underline{X}^{op},\underline{S})$. If X and X' are $\underline{Th}$ objects, a map $f: X \longrightarrow X'$ is a $\underline{Th}$ morphism if for each Y, the induced map $(Y,X) \longrightarrow (Y,X')$ is a $\underline{Th}$ morphism, which means that for each $(n) \rightarrow (m)$ in $\underline{Th}$,

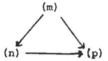

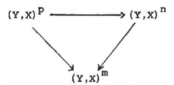

$$\begin{array}{ccc} (Y,X)^m & \xrightarrow{(Y,f)^m} & (Y,X')^m \\ \big\downarrow & & \big\downarrow \\ (Y,X)^n & \xrightarrow{(Y,f)^n} & (Y,X')^n \end{array}$$

commutes. Evidently (using the fact that $\underline{X} \longrightarrow \mathcal{F}(\underline{X}^{op},\underline{S})$ is full and faithful) this is the same as a natural transformation $(-,X) \xrightarrow{\varphi} (-,X')$ such that there is a commutative diagram

$$\begin{array}{ccc} (-,X)^m & \xrightarrow{\varphi^m} & (-,X')^m \\ \big\downarrow & & \big\downarrow \\ (-,X)^n & \xrightarrow{\varphi^n} & (-,X')^n \end{array}$$

corresponding to each $(n) \longrightarrow (m)$ in $\underline{Th}$.

(5.11) Theorem. Let $\underline{X}$ be regular (resp. exact) and $\underline{Th}$ be a finitary theory. Then $\underline{X}^{\underline{Th}}$ is also regular (resp. exact). The underlying $\underline{X}^{\underline{Th}} \longrightarrow \underline{X}$ is a reflexively exact functor.

Proof. It is clear that $\underline{X}^{\underline{Th}} \longrightarrow \underline{X}$ creates all inverse limits which exist in $\underline{X}$ and in particular reflects isomorphisms. The above discussion shows that it is sufficient to consider the case that $\underline{X}$ has finite products. Now suppose that

$$X' \rightrightarrows X \longrightarrow X''$$

is exact in $\underline{X}$ and that X' and X have been equipped with $\underline{Th}$ structures in such a way that $X' \rightrightarrows X$ are morphisms of $\underline{Th}$-algebras (i.e. natural transformations). In that case we have an exact sequence, in particular a coequalizer

$$X'^n \rightrightarrows X^n \longrightarrow X''^n,$$

and corresponding to any map $(1) \longrightarrow (n)$ in $\underline{Th}$ there is a commutative

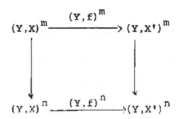

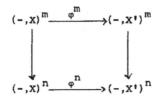

diagram

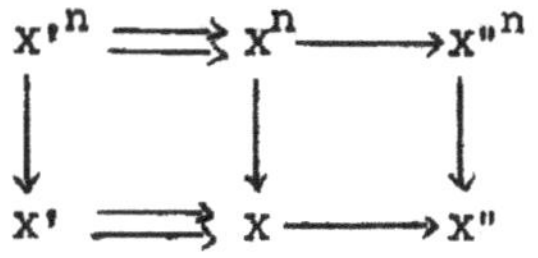

the right hand arrow being induced by the coequalizer. This induces all the operations on X" in such a way that $X \longrightarrow X"$ is a map of algebras as soon as we know that X" is an algebra, i.e. satisfies the equations. To show that, take a commutative triangle

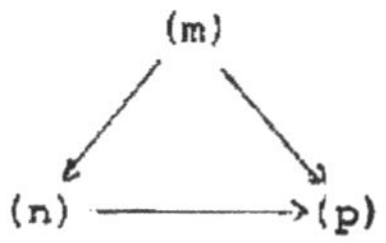

in $\underline{Th}$ and consider

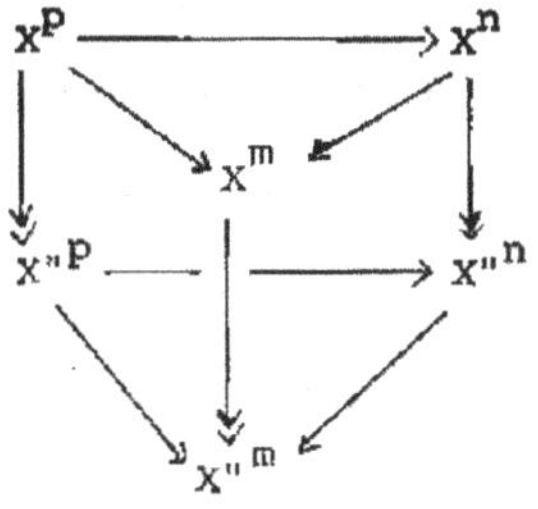

in which each vertical square and the top triangle commute. Since $X^p \longrightarrow\!\!\!\!\rightarrow X"^p$, this can be canceled to show that the bottom triangle $\underline{X}^{\underline{Th}} \longrightarrow X$ creates $\longrightarrow\!\!\!\!\rightarrow$ and hence is exact. In particular, starting with

in $\underline{X}^{\underline{Th}}$, we can pull it back in $\underline{X}$, and the pullback will automatically be an $\underline{X}^{\underline{Th}}$ algebra and the maps $\underline{X}^{\underline{Th}}$ morphisms. The appropriate arrow will be $\longrightarrow\!\!\!\!\rightarrow$ in $\underline{X}$, and by the above in $\underline{X}^{\underline{Th}}$ as well. Now suppose that $\underline{X}$ is exact. Given $X' \rightrightarrows X$ in $\underline{X}^{\underline{Th}}$, which is an equivalence relation in $\underline{X}^{\underline{Th}}$, then it follows from (5.6) that it is an equivalence relation in $\underline{X}$ as well. But then it is part of an exact sequence in $\underline{X}$ and the third term can be given a unique $\underline{Th}$ structure so that it is exact in $\underline{X}^{\underline{Th}}$ as well.

(5.12) Theorem. Let $U: \underline{X} \longrightarrow \underline{Y}$ be an exact functor and $\underline{Th}$ a finitary theory. Then there is a natural lifting $U^{\underline{Th}}: \underline{X}^{\underline{Th}} \longrightarrow \underline{Y}^{\underline{Th}}$ such that

$$\begin{array}{ccc} \underline{X}^{\underline{Th}} & \xrightarrow{\quad U^{\underline{Th}} \quad} & \underline{Y}^{\underline{Th}} \\ \downarrow & & \downarrow \\ \underline{X} & \xrightarrow{\quad U \quad} & \underline{Y} \end{array}$$

is commutative. Moreover $U^{\underline{Th}}$ is exact.

Proof. Except for the last line, this is an easy consequence for any U which preserves finite products. The last assertion is also easy, since the other functors in the diagram are exact and $\underline{Y}^{\underline{Th}} \longrightarrow \underline{Y}$ is reflexively exact.

(5.13) Remark. When $\underline{X} = \underline{S}$, (5.11) is true for all theories $\underline{Th}$ (not just finitary ones). This can be easily proved (by the same argument) for any $\underline{X}$ which satisfies the following. The n-th power functor exists and is exact for all cardinal numbers n. For this we need only that n-th powers exist and preserve $\longrightarrow\!\!\!\!\rightarrow$. Or these conditions may be valid for all $n < N_o$. In that case, the result bolds for all theories $\underline{Th}$ of rank $< N_o$. Similar remarks apply to (5.12) when $\underline{X}$ and $\underline{Y}$ have, and U preserves all n-th powers, or n-th powers for all $n < N_o$, as the case may be.

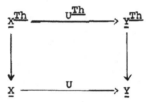

Chapter II. Locally Presentable Categories.

1. Definitions.

(1.1) What follows here is a brief description of a more general theory due to Gabriel and Ulmer, as yet unpublished (except as an outline [Ul]) Some of the definitions here differ slightly from theirs in that I restrict consideration to colimits of monomorphic families. I rather think that for exact categories this does not really give a more general theory, although the cardinal numbers used to satisfy some of the definitions might become larger. Throughout this chapter, $\underline{X}$ and $\underline{Y}$ will be two regular categories which are cocomplete.

(1.2) Definition. Let $\underline{I}$ be a partially ordered set and n be a cardinal number. We say that $\underline{I}$ is $\leq n$ directed if every set of $\leq n$ elements of $\underline{I}$ has an upper bound in $\underline{I}$. An n-filter in $\underline{X}$ is a functor $D: \underline{I} \to \underline{X}$ with $\underline{I}$ $\leq n$ directed and such that for each $i \leq j$ in $\underline{I}$, the value of D at $i \longrightarrow j$, denoted $D(j,i)$, is a monomorphism. Sometimes, for emphasis we will call it a mono-filter. An object $X \in \underline{X}$ is said to have rank $\leq n$ if for every n-filter $D: \underline{I} \longrightarrow \underline{X}$, $(X, \operatorname{colim} Di) \xrightarrow{\simeq} \operatorname{colim}(X,Di)$.

(1.3) Definition. A set Γ of objects of $\underline{X}$ is said to be a set of generators of $\underline{X}$ if for every $f: X \rightarrowtail X'$ which is not an isomorphism there is a $G \in \Gamma$ and a map $G \longrightarrow X'$ which does not factor through f. $\underline{X}$ is said to be locally presentable if it has arbitrary coproducts (denoted $\coprod$) and a set of generators each one of which has rank.

(1.4) Proposition. Let $\underline{X}$ be locally presentable. Then for any $X \in \underline{X}$, there is a $\coprod_{j \in J} G_j \twoheadrightarrow X$ where, for each $j \in J$, $G_j \in \Gamma$.

Proof. Form $\coprod_{G \in \Gamma} \coprod_{(G,X)} G$, the coproduct of one copy of G for each map to X from each $G \in \Gamma$. There is a canonical evaluation $e: \coprod \coprod G \longrightarrow X$ defined by $e.\langle u\rangle = u$ where $\langle u\rangle : G \longrightarrow \coprod \coprod G$ is the coordinate injection corresponding to $u\colon G \longrightarrow X$. Factor e as

$$\coprod\coprod G \xrightarrow{e_o} X_o \rightarrowtail^{f} X.$$

If $u\colon G \longrightarrow X$ is any map, $e.\langle u\rangle = u$ so that $u = f.e_o.\langle u\rangle$ factors through f. Since this is true for all such u, f must be an isomorphism.

(1.5) It is easy to see that the above characterization could have been taken as the definition of this kind of generator. To distinguish it from the more common kind of generator, whose definition is equivalent (in the presence of coproducts) to the same map being an ordinary epimorphism, these could be called a set of regular generators. Here, however, we will simply call them generators.

(1.6) Proposition. Let $f\colon X \longrightarrow X'$. Then

a) If (G,f) is $\twoheadrightarrow$ for all $G \in \Gamma$, f is $\twoheadrightarrow$.

b) (G,f) is $\rightarrowtail$ for all $G \in \Gamma$ if and only if f is $\rightarrowtail$.

c) (G,f) is $\xrightarrow{\simeq}$ for all $G \in \Gamma$ if and only if f is $\xrightarrow{\simeq}$.

Proof. a) This follows easily from

$$\begin{array}{ccc} \coprod\coprod(G,X) & \twoheadrightarrow & \coprod\coprod(G,X') \\ \downarrow & & \downarrow \\ X & \longrightarrow & X' \end{array}$$

b) One way is trivial. If (G,f) is $\rightarrowtail$, consider the diagram

$$X''' \xrightarrow{d} X'' \underset{d^1}{\overset{d^o}{\rightrightarrows}} X \xrightarrow{f} X'$$

in which d^o and d^1 are the kernel pair of f and d is their equalizer, Since $(G,-)$ preserves limits and (G,f) is $\rightarrowtail$, it follows that $(G,d^o) = (G,d^1)$, and then (G,d) is an isomorphism. Since d is a monomorphism, it follows from the definition of generator that d is

$\xrightarrow{\sim}$. But then $d^{o} = d^{1}$, which in turn implies that f is $\rightarrowtail$.

c) This is now clear.

(1.7) Remark. It is clear from the above argument that, in particular, the more usual definition of generator is also satisfied.

2. Preliminary results.

Throughout this section X is a cocomplete regular category and Γ a set of generators.

(2.1) Proposition. X is well-powered.

Proof. For any object X a subobject X_o is determined by those maps from a G ∈ Γ which factor trough X_o. In other words, there are no more subobjects of X than there are subsets of ∪(G,X), the union taken over G ∈ Γ.*

(2.2) Corollary. Each object of X has only a set of regular quotients.

Proof. A regular quotient of X is determined by its kernel pair, and that is a subobject of X × X.

(2.3) Proposition. Let D: I ⟶ X be a small diagram. Then the set (Γ,D) of all objects (G,γ) ∈ (X,D) for which G ∈ Γ form a generating set in (X,D).

Proof. It is a set since each G has only a set of maps to a small diagram. If $X \overset{f}{\rightarrowtail} Y \longrightarrow D$ is a monomorphism, not an isomorphism in (X,D), then $X \overset{f}{\rightarrowtail} Y$ is a monomorphism as noted in I, ((5.3) above) and clearly not an isomorphism, as the inverse would also be a map of (X,D). Then there is a map G ⟶ Y which does not factor through X, and if we use the composite G ⟶ Y ⟶ D to lift G into (X,D) it becomes an element of (Γ,D) with the required property.

(2.4) Theorem. Let X be a cocomplete, regular category with a set of regular generators and such that each object has only a set of regular quotients. Then X is complete.

Proof. For a diagram D: I ⟶ X, a limit of D is a terminal object of (X,D). It is easily seen that cocompleteness is inherited by that

*For nested subobjects, this is clear from the definition of generator. For others, consider the intersection and reduce to the previous case.

category as well as the property of each object having a set of regular quotients. By I. (5.4) and (2.3) the other properties of the statement are also inherited. Hence it suffices to show that such an $\underline{X}$ always has a terminal object. Let Γ be the set of generators, $X = \coprod G$, $G \in \Gamma$, and Q be the colimit of all the regular quotients of X. First I claim that Q is itself a regular quotient of X. It is sufficient to show that every commutative square

$$\begin{array}{ccc} X & \longrightarrow & Q \\ \downarrow & & \downarrow \\ Y & \rightarrowtail & Z \end{array}$$

has a diagonal fill-in. (Just take $Z = Q$ and Y the image of X in Q.) But by commutativity of the diagram, we have, for each regular quotient $X \twoheadrightarrow X'$,

$$\begin{array}{ccc} X & \twoheadrightarrow & X' \\ \downarrow & \swarrow & \downarrow \\ & & Q \\ & & \downarrow \\ Y & \rightarrowtail & Z \end{array},$$

giving a family $X' \longrightarrow Y$, obviously coherent and extending to $Q \longrightarrow Y$. Thus Q itself can have no regular quotient, for that would be a further regular quotient of X. For any $Y \in \underline{X}$, there will be a map $\coprod G_i \twoheadrightarrow Y$, and evidently there is a $\coprod G_i \longrightarrow X$, since X is the coproduct of all the $G \in \Gamma$. Pushing out, we get

$$\begin{array}{ccccc} \coprod G_i & \longrightarrow & X & \longrightarrow & Q \\ \downarrow\!\downarrow & & & & \downarrow\!\downarrow \\ Y & & \longrightarrow & & Q' \end{array}$$

whence $Q \cong Q'$ and $(Y,Q) \neq \emptyset$. If there were distinct maps $Y \rightrightarrows Q$ for some Y, their coequalizer would be a regular quotient of Q.

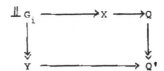

(2.5) Remark. It should be noted that this method works for any factorization system and is a form of the special adjoint functor theorem. That is, if there is some factorization system and generators such that the appropriate map is an epimorphism for that system, and if the objects have only a set of quotients in that system, then the special adjoint functor theorem (here in dual form) holds.

(2.6) Proposition. Suppose $\underline{I}$ is some index category; $D: \underline{I} \longrightarrow \underline{X}$, $E: \underline{I} \longrightarrow \underline{X}$ are functors; and $D \longrightarrow E$ is a natural transformation such that $D_i \twoheadrightarrow E_i$ for all i. Then $\text{colim } D \twoheadrightarrow \text{colim } E$.

Proof. Let $X = \text{colim } D_i$, $Y = \text{colim } E_i$. For each i we have a commutative diagram

$$\begin{array}{ccccc} D_i \times_{E_i} D_i & \overset{d_i^0}{\underset{d_i^1}{\rightrightarrows}} & D_i & \twoheadrightarrow & E_i \\ \downarrow & & \downarrow & & \downarrow \\ X \times_Y X & \overset{d^0}{\underset{d^1}{\rightrightarrows}} & X & \longrightarrow & Y \end{array}$$

Given $X \longrightarrow Z$, which coequalizes d^0, d^1, this induces $E_i \longrightarrow Z$, which coequalizes d_i^0 and d_i^1 and induces a unique $E_i \longrightarrow Z$ making the diagram commute. This family of maps is easily seen to be natural in i, and then there is further induced a map $Y \longrightarrow Z$. Then the outer pentagon of

$$\begin{array}{ccc} D_i & \longrightarrow & E_i \\ \downarrow & & \downarrow \\ X & \longrightarrow & Y \\ & \searrow \quad \swarrow & \\ & Z & \end{array}$$

commutes for each i. Since $X = \text{colim } D_i$, this implies that the triangle commutes.

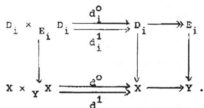

3. Rank.

(3.1) Throughout this section, $\underline{X}$ will denote a locally presentable regular category and Γ a set of generators with rank. We will suppose that n_1 is an infinite cardinal number sufficiently large that $n_1 \geq \#(\Gamma)$ (# is used to denote cardinality) and $n_1 \geq$ the rank of every object of Γ.

(3.2) Let Γ_1 denote the set of coproducts of n_1 or fewer objects of Γ and Γ_2 denote the set of regular quotients of objects of Γ_1. Let $n_2 = \sup_{X \in \Gamma_2} \#(\bigcup_{G \in \Gamma} (G,X))$ and $n = 2^{2^{n_2}}$. Let $\underline{X}_n$ denote the full subcategory of $\underline{X}$ consisting of all objects whose rank $\leq n$.

(3.3) Proposition. With n and $\underline{X}_n$ as above, the objects $X \in \underline{X}_n$ are characterized by each of the following properties.

a) There is a map $\coprod_{i \in I} G_i \twoheadrightarrow X$ with each $G_i \in \Gamma$ and such $\#(I) \leq n$.

b) $\#(\bigcup_{G \in \Gamma} (G,X)) \leq n$.

This remains true for any power cardinal $\geq n$.

Before giving the proof, we require the following.

(3.4) Proposition. Every object of $\underline{X}$ is a colimit of those subobjects of it which satisfy condition a).

Proof. Let $X \in \underline{X}$ and consider the set of all subobjects of X which satisfy condition a). It follows from (2.6) that the objects satisfying condition a) are closed under n-fold coproducts and, by forming images, that these subobjects form an n-filter. Let X' be its colimit. For $G \in \Gamma$, any map $G \longrightarrow X$ lands in a subobject of X satisfying a), namely its image, and hence factors through X'. Thus $(G,X') \twoheadrightarrow (G,X)$. If two different maps $G \rightrightarrows X'$ are given, each of them, since rank $G \leq n_1 < n$, must factor through one of the given subobjects of X and, by directedness, through some one subobject. Thus, since they factor through a subobject of X, they must remain distinct in X. Thus $(G,X') \rightarrowtail (G,X)$ also, and by (1.6) $X' \xrightarrow{\simeq} X$.

(3.5) Proof of (3.3). Write $X = \operatorname{colim} X_j$ where X_j ranges over the subobject of X satisfying condition a). Now since rank $X \leq n$, the identity map $X \longrightarrow X$, being a map to the colimit of an n-filter, must factor through one of the objects in that filter. This evidently implies that X itself is one of them and so satisfies a). Now suppose an object satisfies a). Then for each $J \subset I$ such that $\#(J) \leq n_1$, let X_J be the image $\coprod_{i \in J} G_i \longrightarrow X$. Then evidently $X_J \in \Gamma_2$, and so $\#(\bigcup_{G \in \Gamma} (G,X_J)) \leq n_2$. The number of such subsets of I is limited by $n^{n_1} = (2^{n_2})^{n_1} = 2^{n_2 \times n_1} = 2^{n_2} = n$. It is clear that the set of all X_J is an n_1-filter on X. Just as above, this permits showing that for each $G \in \Gamma$, $(G, \operatorname{colim} X_J) \rightarrowtail (G,X)$, and hence by (1.6) that $\operatorname{colim} X_J \rightarrowtail X$. On the other hand, each of the $G_i \longrightarrow X$ factors through one of the X_J, and hence we have a factorization

$$\coprod G_i \longrightarrow \operatorname{colim} X_J \longrightarrow X$$

whose composition is $\twoheadrightarrow$, which shows that the second factor is also. Thus $X = \operatorname{colim} X_J$. Now $(G, \operatorname{colim} X_J) = \operatorname{colim}(G,X_J)$, and so

$$\#(\bigcup_{G \in \Gamma} (G,X)) = \#(\bigcup_{G \in \Gamma} \operatorname{colim}(G,X_J))$$

$$\leq \sum_{G \in \Gamma} \#(\operatorname{colim}(G,X_J)) \leq \sum_{G \in \Gamma} \sum_{J \in I} \#(G,X_J)$$

$\leq n_1 \cdot n \cdot n_2 = n$. Thus condition a) implies condition b) and the reverse implication is obvious. Now suppose an object X satisfies condition a) and we have an n-filter $\{Y_j | j \in I\}$. We see from $(G,Y_j) \rightarrowtail \operatorname{colim} (G,Y_j) \xrightarrow{\simeq} (G, \operatorname{colim} Y_j)$ and (1.6) that $Y_j \rightarrowtail \operatorname{colim} Y_j$. Now supposing $\coprod_{i \in I} G_i \twoheadrightarrow X$ and $\#(I) \leq n$, we use the readily proved fact that in $\underline{S}$, I-indexed products commute with n-filters and thus

$$(\coprod G_i,\ \mathrm{colim}\ Y_j) \simeq \Pi(G_i,\ \mathrm{colim}\ Y_j)$$
$$\simeq \Pi\ \mathrm{colim}(G_i,Y_j) \simeq \mathrm{colim}\ \Pi(G_i,Y_j)$$
$$\simeq \mathrm{colim}(\Pi G_i,Y_j),$$ which shows that ΠG_i has rank $\leqslant n$.

The fact that X does follows from a diagonal fill-in in the diagram

$$\begin{array}{ccc} \coprod G_i & \longrightarrow\!\!\!\!\!\!\rightarrow & X \\ \downarrow & & \downarrow \\ Y_j & \rightarrowtail & \mathrm{colim}\ Y_j . \end{array}$$

The last remark about power cardinals $\geqslant n$ is trivial from the proof.

(3.6) <u>Corollary</u>. $\underline{X}_n$ is n-cocomplete, finitely complete, and closed under sub- and regular quotient objects.

Proof. It is clear that the condition a) above is inherited by n-fold coproducts as well as by regular quotients while condition b) is inherited by subobjects and finite products (in fact, by n_2-fold products).

(3.7) <u>Corollary</u>. Every object of $\underline{X}$ is the colimit of those subobjects of it which belong to $\underline{X}_n$.

(3.8) <u>Corollary</u>. $\underline{X}_n$ is a dense subcategory of $\underline{X}$.

Proof. This means that every $X \in \underline{X}$ is the colimit of the functor $(\underline{X}_n, X) \longrightarrow \underline{X}$ which associates to each $X' \longrightarrow X$ the domain X'. By factoring every such map as $\cdot \twoheadrightarrow \cdot \rightarrowtail \cdot$ and using the fact that $\underline{X}_n$ is closed under regular quotients, we see that the monomorphisms in $(\underline{X}_n, X)$ are cofinal. Thus the colimits are the same and the result is a corollary of (3.7).

(3.9) <u>Proposition.</u> Let $X \in \underline{X}$ and $X' \in \underline{X}_n$. Given any $X \twoheadrightarrow X'$, there is an $\underline{X}_n$ subobject $X'' \rightarrowtail X$ such that the composite $X'' \rightarrowtail X \twoheadrightarrow X'$ is $\twoheadrightarrow$.

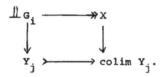

Proof. Consider a map $\coprod_{i \in I} G_i \longrightarrow\!\!\!\!\rightarrow X$. Among all the composites $G_i \longrightarrow \coprod G_i \longrightarrow\!\!\!\!\rightarrow X \longrightarrow X'$ there can be at most n distinct maps. Choose $J \subset I$ so that the set of such composite maps for $i \in J$ is represented exactly once for $i \in J$. Then $\#(J) \leq n$, while evidently $\coprod_{i \in J} G_i \to X \longrightarrow X'$ must have the same image in X' and hence is $\longrightarrow\!\!\!\!\rightarrow$. Then let X'' be the image of $\coprod_{i \in J} G_i \longrightarrow X$.

4. Kan extension of functors.

The purpose of this section is to prove:

(4.1) Theorem: Let $\underline{X}$ and $\underline{Y}$ be locally presentable regular categories and n be a cardinal such that $\underline{X}_n$ satisfies (3.3) and such that $\underline{Y}_n$ contains a set of generators of $\underline{Y}$. Suppose $U: \underline{X}_n \longrightarrow \underline{Y}_n$ is a functor and let $\tilde{U}: \underline{X} \longrightarrow \underline{Y}$ be its Kan extension. Then:

a) If U is reflexively exact, so is $\tilde{U}$.

b) If U is faithful (resp. full and faithful), so is $\tilde{U}$.

(4.2) The rest of this section is devoted to proving this theorem. Without further mention, $\underline{X}$, $\underline{Y}$, n, U, and $\tilde{U}$ will be as in the statement.

(4.3) Proposition. Colimits of n-filters in $\underline{Y}$ commute with finite limits.

Proof. Suppose we are given n-filters $\{Y_i'\}$ and $\{Y_j''\}$ indexed by $i \in I$, $j \in J$, and we let $Y' = \text{colim } Y_i'$, $Y'' = \text{colim } Y_i''$, $Y_{ij} = Y_i' \times Y_j''$, and $Y = \text{colim } Y_{ij}$. Then we want to show that the natural map $Y \xrightarrow{\sim} Y' \times Y''$. We use (1.6) Let Λ be a generating set in $\underline{Y}_n$. For $L \in \Lambda$, $(L,Y) \cong (L,\text{colim } Y_{ij}) \cong \text{colim}(L,Y_{ij}) \cong \text{colim}(L,Y_i' \times Y_j'') \cong \text{colim}((L,Y_i')\times(L,Y_j'')) \cong \text{colim}(L,Y_i') \times \text{colim}(L,Y_j'')$. (since directed colimits commute with finite limits in $\underline{S}$) $\cong (L,\text{colim } Y_i')\times(L,\text{colim } Y_j'') \cong (L,Y')\times(L,Y'') \cong (L,Y' \times Y'')$. The proof for equalizers is similar and we omit it. It is not necessary to have, in that case, maps $Y_i' \rightrightarrows Y_j''$ given for all i,j but only for sufficiently many pairs of indices that the resulting subset of $I \times J$ remain n-directed.

(4.4) Proposition. Let $X', X'' \in \underline{X}$. Then the set of maps $X_i' \times X_j'' \rightarrowtail X' \times X''$, indexed by all $\underline{X}_n$ subobjects $X_i' \rightarrowtail X'$ and all $\underline{X}_n$ subobjects $X_j'' \rightarrowtail X''$, is cofinal among all the $\underline{X}_n$-subobjects of $X' \times X''$.

Proof. Given $X_k \rightarrowtail X' \times X''$ with $X_k \in \underline{X}_n$, we let X_k' be the image of $X_k \longrightarrow X' \times X'' \longrightarrow X'$ and similarly X_k'' the image in X". Then, since products of $\rightarrowtail$ are certainly $\rightarrowtail$, and from the universal mapping property of products, we have

$$X_k \rightarrowtail X_k' \times X_k'' \rightarrowtail X' \times X''.$$

(4.5) Proposition. Let $X' \longrightarrow X \rightrightarrows X''$ be an equalizer diagram in $\underline{X}$. Then each $\underline{X}_n$ subobject $X_i' \rightarrowtail X'$ appears at least once among the possible equalizer diagrams

$$X_i' \longrightarrow X_j \rightrightarrows X_k''$$

in which X_j and X_k'' are $\underline{X}_n$ subobjects of X and X" respectively.

Proof. Let $X_j = X_j'$ itself and X_k'' be the image in X" of the equal maps

$$X_i' \longrightarrow X' \rightrightarrows X''.$$

(4.6) Remark. The implication of these last two propositions is that for $X = X' \times X''$, the functor which associates to $X_i' \rightarrowtail X'$ and $X_j'' \rightarrowtail X''$, $X_i' \times X_j'' \rightarrowtail X' \times X''$ is cofinal. Similarly, suppose $X' \longrightarrow X \rightrightarrows X''$ is an equalizer diagram. Then the functor which, to each pair $X_j \rightarrowtail X$, $X_k'' \rightarrowtail X''$ for which the restrictions take X_j into X_k'', associates the equalizer of these restrictions is cofinal.

(4.7) Proposition. Given $X \rightrightarrows X''$ as above, let $\{X_j \mid j \in J\}$ and $\{X_k'' \mid k \in K\}$ be the n-filters of $\underline{X}_n$ subobjects of X and X" respectively. Let L be the subset of $J \times K$ of those pairs (j,k) for which the restrictions of the given maps each take X_j into X_k''. Then L is an n-directed set.

Proof. Given n or fewer indices of L, we can find j greater than any of the first coordinates and k' greater than any of the second. We have morphisms

$$\begin{array}{l} X_j \searrow \\ X_j \longrightarrow X'' \\ X''_{k'} \nearrow \end{array}$$

where X_j and $X''_{k'}$ both belong to $\underline{X}_n$. Let + denote coproduct and X_k be the image of $X_j + X_j + X''_{k'} \longrightarrow X''$. Clearly the domain of that map belongs to $\underline{X}_n$ and $(j,k) \in L$ dominates each of the given indices.

(4.8) Corolloary. If U preserves finite limits, so does $\tilde{U}$.

(4.9) Proposition. If U preserves $\longrightarrow\!\!\!\!\rightarrow$, so does $\tilde{U}$.

Proof. Let $X \twoheadrightarrow X'$. For any $\underline{X}_n$ subobject $X'_o \rightarrowtail X'$, we pull back to get

$$\begin{array}{ccc} X_1 & \twoheadrightarrow & X'_o \\ \downarrow & & \downarrow \\ X & \twoheadrightarrow & X' \end{array}$$

and let $X_o \rightarrowtail X_1$, be an $\underline{X}_n$ subobject, whose existence is guaranteed by (3.9), such that $X_o \twoheadrightarrow X'_o$. Then $UX_o \twoheadrightarrow UX'_o$. Now if I and J are the index sets for the $\underline{X}_n$-subobjects of X and X' respectively, what we have is a map $j \longmapsto i(j)$ of $J \longrightarrow I$ such that $X_{i(j)} \twoheadrightarrow X'_i$. Then $\mathrm{colim}\, UX_{i(j)} \longrightarrow \mathrm{colim}\, UX_i \longrightarrow \mathrm{colim}\, UX'_j$ is such that the composite is $\twoheadrightarrow$ by (2.6). This implies that the second is also. This second map is just $\tilde{U}X \twoheadrightarrow \tilde{U}X'$.

(4.10) Proposition. If U reflects monomorphisms, so does $\tilde{U}$.

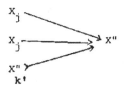

Proof. Let $f\colon X \longrightarrow X'$ be a map such that $Uf\colon \tilde{U}X \rightarrowtail \tilde{U}X'$. If f is not $\rightarrowtail$, then there are two maps $X'' \overset{d^o}{\underset{d^1}{\rightrightarrows}} X \longrightarrow X'$ which are co-equalized by f and, as observed in (1.7), there is a $G \in \Gamma$ and a map $G \longrightarrow X''$ which does not equalize d^o and d^1. Let X''_o be the image of G in X" and X_o be the image of $G + G \longrightarrow X$. Then we have

$$\begin{array}{ccccc} X_o'' & \overset{e^o}{\underset{e^1}{\rightrightarrows}} & X_o & & \\ \downarrow & & \downarrow & & \\ X'' & \overset{d^o}{\underset{d^1}{\rightrightarrows}} & X & \xrightarrow{f} & X' \end{array}$$

with X_o'' and X_o in $\underline{X}_n$ and $e^o \neq e^1$. Now apply $\tilde{U}$ to get

$$\begin{array}{ccccc} UX_o'' & \overset{Ue^o}{\underset{Ue^1}{\rightrightarrows}} & UX_o & & \\ \downarrow & & \downarrow & & \\ \tilde{U}X'' & \overset{\tilde{U}d^o}{\underset{\tilde{U}d^1}{\rightrightarrows}} & \tilde{U}X & \xrightarrow{\tilde{U}f} & \tilde{U}X'. \end{array}$$

Now U reflects isomorphisms and is faithful, so that $Ue^o \neq Ue^1$, which implies that $Ud^o \neq Ud^1$; while $Uf.Ud^o = Uf.Ud^1$ contradicts Uf being $\rightarrowtail$.

(4.11) Proposition. If U reflects isomorphisms, so does $\tilde{U}$.

Proof. First I claim that U reflects $\rightarrowtail$. If $f: X \longrightarrow X'$ is such that $Ug: UX \rightarrowtail UX'$, consider

$$X''' \longrightarrow X'' \rightrightarrows X \xrightarrow{f} X'$$

where $X'' \rightrightarrows X$ is the kernel pair of f and $X''' \longrightarrow X''$ is the equalizer of them. Apply U and reason as in the proof of (1.6). Now suppose that $\tilde{U}f: \tilde{U}X \xrightarrow{\sim} \tilde{U}X'$. By (4.10), $f: X \rightarrowtail X'$. If this is not an $\xrightarrow{\sim}$, there is a map $G \longrightarrow X'$ which does not factor through f. If we let X_o' be the image of $G \longrightarrow X'$ and X_o be the pullback in

$$\begin{array}{ccc} X_o & \overset{f_o}{\rightarrowtail} & X'_o \\ \downarrow & & \downarrow \\ X & \overset{f}{\rightarrowtail} & X', \end{array}$$

it is clear that $X'_o \in \underline{X}_n$, and X_o, being a subobject of $\underline{X}'$, is also. Now apply $\tilde{U}$ to get the diagram

$$\begin{array}{ccc} UX_o & \overset{Uf_o}{\rightarrowtail} & UX'_o \\ \downarrow & & \downarrow \\ \tilde{U}X & \overset{\tilde{U}f}{\rightarrowtail} & \tilde{U}X'. \end{array}$$

If $\tilde{U}f$ is an isomorphism, so is Uf_o, since the diagram remains a pull-back; and then $f_o \colon X_o \xrightarrow{\simeq} X'_o$. But this implies that the given map $G \longrightarrow X'$ really does factor through f, and we have a contradiction.

(4.12) Proposition. Let U be faithful (resp. full and faithful). Then $\tilde{U}$ is also.

Proof. Write $X = \operatorname{colim} X_i$, $X' = \operatorname{colim} X'_j$, each colim taken over the diagram of $\underline{X}_n$ subobjects of X and X' respectively. Of course from the properties of $\underline{X}_n$ it is clear that these diagrams are n-directed. Then

$$(X,X') \cong (\operatorname{colim} X_i, \operatorname{colim} X'_j) \cong \lim(X_i, \operatorname{colim} X'_j) \underset{(1)}{\cong} \lim \operatorname{colim} (X_i, X'_j)$$
$$\underset{(2)}{\cong} \lim \operatorname{colim} (UX_i, UX'_j) \underset{(3)}{\cong} \lim (UX_i, \operatorname{colim} UX'_j) \cong$$
$$\cong (\operatorname{colim} UX_i, \operatorname{colim} UX'_j) \cong (\tilde{U}X, \tilde{U}X').$$

The arrows labeled (1) and (3) are isomorphisms because X_i and UX_i are objects of rank $\leq n$ in $\underline{X}$ and $\underline{Y}$ respectively. If U is faithful (resp. full and faithful), then the arrow labeled (2) is for each i and j a monomorphism (resp. isomorphism) and both directed colimit and arbitrary limit preserve monomorphisms, while, of course, everything preserves isomorphisms. Hence $\tilde{U}$ will also be faithful (resp. full and faithful).

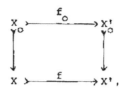

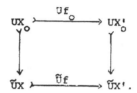

5. Toposes.

(5.1) We have already seen how every small regular category has a full exact embedding into a topos. Moreover, every regular category has a full exact embedding into an illegimate topos. In this section we will show that every cocomplete locally presentable exact category has a full exact embedding into a topos, while, conversely, a topos is itself a locally presentable exact category. We begin with the latter.

(5.2) Theorem: Every topos is locally presentable.

Proof. Let $\underline{E}$ be a topos, and write $\underline{E} = \mathfrak{F}(\underline{C}^{op},\underline{S})$ for some small category $\underline{C}$ and some topology on $\underline{C}$ which is less fine than the canonical topology. Let n be an infinite cardinal number sufficiently large that no covering in the topology on $\underline{C}$ has more than n-elements. Then, as is well known, the objects of $\underline{C}$ (i.e. the representable functors) form a set of generators. I claim that each $C \in \underline{C}$ has rank $\leq n$ in $\underline{E}$. Since in the whole functor category, $(-,C)$ commutes with all colimits (by the Yoneda lemma, $((-,C), \operatorname{colim} G_i) = \operatorname{colim} G_iC = \operatorname{colim}((-,C),G_i))$, it is sufficient to show that if $D: \underline{I} \longrightarrow \underline{E}$ is a functor with $\underline{I}$ an n-directed index set, then the colim D_i is the same in $\underline{E}$ as in $(\underline{C}^{op},\underline{S})$; or, which is the same thing, to show that an n-directed colimit of sheaves is a sheaf. So suppose $\{C_j \longrightarrow C \mid j \in J\}$ is a covering of C and I is an n-directed set. In $\underline{S}$, n-directed colimits commute with $\leq$ n-fold products and, since n is infinite, with equalizers. If $F = \operatorname{colim} D_i$, we have that

$$FC \longrightarrow \Pi FG_j \rightrightarrows \Pi F(C_{j_1} \times_C C_{j_2})$$

is isomorphic to

$$\operatorname{colim} D_i(C) \longrightarrow \Pi \operatorname{colim} Di(C_j) \rightrightarrows \Pi \operatorname{colim} Di(C_{j_1} \times_C C_{j_2})$$

which is isomorphic to

$$\operatorname{colim} Di(C) \longrightarrow \operatorname{colim} \Pi Di(C_j) \rightrightarrows \operatorname{colim}(\Pi Di(C_{j_1} \times_C C_{j_2}))$$

which, since each Di is a sheaf, is a directed colimit of equalizers and again an equalizer.

(5.3) Corollary. Every cocomplete locally presentable regular category has a full exact embedding into a topos.

Proof. Let $\underline{X}$ be such a category and find a cardinal n such that $\underline{X}_n$ satisfies (3.3). Let $\underline{C} = \underline{X}_n$, and we have an embedding of $\underline{X}_n \longrightarrow \mathcal{S}(\underline{C}^{op}, \underline{S})$ which, since the cardinality of each covering of the topology is 1, embeds $\underline{X}_n$ as objects of finite rank. Then the hypotheses of (4.1) are satisfied.

Chapter III. The Embedding

1. Statements of result.

(1.1) Theorem. Every locally presentable category has a full exact embedding into a functor category.

(1.2) Theorem. Every topos has a full exact embedding into a functor category.

(1.3) Theorem. Every small regular category has a full exact embedding into a functor category.

(1.4) Theorem. Every small, finitely complete regular category has a full exact embedding into objects of finite rank of a functor category.

(1.5) Except for the last clause of (1.4), it is clear from I. (4.4) II. (4.1) and II. (5.2) that these statements are all equivalent. That last clause could also be derived from the previous theorems, but since we have to prove something, we will prove (1.4). In fact, we will prove something even stronger. Recall that an object $\emptyset$ of a category is an empty object if it is initial and if every map to it is an isomorphism. Let us denote the terminal object of $\underline{X}$ by 1. Then,

(1.6) Theorem: Let $\underline{X}$ be a small finitely complete regular category. Then there is a small category $\underline{C}$, whose objects may be identified with the non-empty subobjects of 1, and a full exact embedding $\underline{X} \longrightarrow (\underline{C}^{op}, \underline{S})$ which sends each object of X to a regular quotient of a representable functor.

(1.7) Proposition. A regular quotient of a representable functor has finite rank.

Proof. As observed above (in the proof of II. (5.2)), any representable functor has finite rank - its hom commutes with all colimits. If $\{F_i\}$ is a monofilter (cf. II. (1.2)) of functors and $F = \operatorname{colim} F_i$, then for each representable functor $(-,C)$,

$$((-,C),F) = \operatorname{colim}((-,C),F_i).$$

The filter of sets $((-,C),F_i)$ is still a monofilter, which implies that $((-,C),F_i) \rightarrowtail ((-,C),F)$ and by II.(1.6) that $F_i \rightarrowtail F$. Now suppose $E \in (\underline{C}^{op},\underline{S})$ is a regular quotient of $(-,C)$. To see that $\operatorname{colim}(E,E_i) \xrightarrow{\simeq} \operatorname{colim}(E,F)$, first observe that by the above, the natural map is 1-1. To show it is onto, consider a map $E \longrightarrow F$. The composite $(-,C) \longrightarrow E \longrightarrow F$ must factor through some F_i and the result is obtained from the diagram

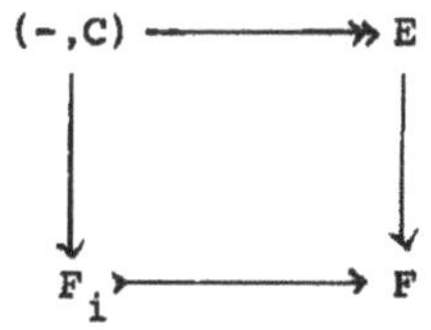

by filling in the diagonal.

(1.8) Corollary. Let $\underline{X}$ be a small, finitely complete regular category in which the terminal object has no non-empty subobject. Then there is a monoid C and full exact embedding $\underline{X} \longrightarrow \underline{S}^C$.

(1.9) Corollary (Mitchell). Let $\underline{A}$ be a small, finitely complete regular additive category (or locally presentable or an $\underline{Ab}$-topos). Then $\underline{A}$ has a full exact embedding into a category of modules.

Proof. Take an embedding into $\underline{S}^C$ as above (there aren't any subobjects of 1 in the additive case). Since it preserves finite products, it lifts to a still exact (additive) embedding into $\underline{Ab}^C$, the category of ZC-modules.

(1.10) The remainder of this chapter is devoted to proving (1.6). Throughout this chapter with the exception of section (2.12)-(2.16),

$\underline{X}$ denotes a small, finitely complete regular category.

2. Support.

(2.1) Choose $X \in \underline{X}$ and factor the terminal map $X \longrightarrow 1$ as $X \twoheadrightarrow S \rightarrowtail 1$. The map $X \longrightarrow S$ is constant, which means that it coequalizes every pair of maps to S. This is because $X \longrightarrow S$ and $X \longrightarrow 1$ have the same kernel pair, $X \times X$. This S is called the support of X and we will write $S = \text{supp } X$.

(2.2) When $\underline{X} = (\underline{C}^{op}, \underline{S})$ and $X \in \underline{X}$, supp X is that functor whose value is 1 wherever the value of X is non-empty and whose value is $\emptyset$ where X's is. Thus supp X is the "characteristic functor" of what would normally be called the support of X.

(2.3) An object $S \in \underline{X}$ will be called a partial terminal object if every map to it is constant.

(2.4) Proposition. Let S be an object of $\underline{X}$. Then the following are equivalent.

a. S is a partial terminal object.

b. The projections p_1, p_2: $S \times S \longrightarrow S$ are equal.

c. The projections p_1, p_2: $S \times S \longrightarrow S$ are equal.

d. The diagonal s: $S \longrightarrow S \times S$ is an isomorphism.

Proof. Trivial.

(2.5) Proposition. Let f: $S \twoheadrightarrow T$ where S is a partial terminal object. Then f is an isomorphism.

Proof. Consider the kernel pair.

(2.6) Proposition. Let f: $X \twoheadrightarrow S$ be constant. Then S is a partial terminal object and $S = \text{supp } X$.

Proof. As any constant map factors through supp X, we have $X \twoheadrightarrow \text{supp } X \twoheadrightarrow S$, the second being $\twoheadrightarrow$ by I (2.5). Now apply

(2.5).

(2.7) Let Supp $\underline{X}$ denote the full subcategory of $\underline{X}$ whose objects are the partial terminal objects. There is at most one map between any two objects of Supp $\underline{X}$ and we will often write $S \leq S'$ for $S \longrightarrow S'$.

(2.8) <u>Proposition</u>. supp: $\underline{X} \to$ Supp$\underline{X}$ is left adjoint to inclusion.

Proof. We must show that for $S \in$ Supp $\underline{X}$, $(X,S) \neq \emptyset$ if and only if $(\text{supp } X, S) \neq \emptyset$. The "if" part is clear from the map $X \longrightarrow \text{supp } X$. and the other follows from the fact that any constant map from X factors through supp X.

(2.9) <u>Proposition</u>. The functor supp preserves finite products.

Proof. Since $X \twoheadrightarrow \text{supp } X \rightarrowtail 1$ and $Y \twoheadrightarrow \text{supp } Y \rightarrowtail 1$, we have, by (2.14),

$$X \times Y \twoheadrightarrow \text{supp } X \times \text{supp } Y \rightarrowtail 1 \times 1 = 1.$$

Thus supp X × supp Y enjoys the characteristic property of supp(X×Y).

(2.10) <u>Proposition</u>. Let X and Y be objects of $\underline{X}$. Then supp X = = supp Y if and only if there is an object Z and maps $Y \twoheadleftarrow Z \twoheadrightarrow X$.

Proof. Given such maps, we conclude from $Z \twoheadrightarrow X \twoheadrightarrow \text{supp } X$ that supp Z = supp X and similarly supp Z = supp Y. Conversely, given supp X = supp Y = S we have

$$\begin{array}{ccc} X \times Y & \twoheadrightarrow & Y \\ \downarrow\!\!\downarrow & & \downarrow\!\!\downarrow \\ Y & \twoheadrightarrow & S \end{array}$$

(2.11) <u>Proposition</u>. Let $\underline{X}$ be regular, $X \in \underline{X}$. $X \times -: \underline{X} \longrightarrow \underline{X}$ reflects isomorphisms if and only if supp X is a terminal object of $\underline{X}$.

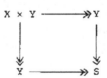

Proof. First observe that $X \times \operatorname{supp} X \longrightarrow X$ by product projection is an isomorphism, since each map to X induces a unique map to supp X. For each $S \in \operatorname{Supp} \underline{X}$, $\operatorname{supp} X \times S = \operatorname{supp}(X \times S)$. Moreover $S \times \operatorname{supp} X \longrightarrow S$ gives $X \times \operatorname{supp} X \times S \longrightarrow X \times S$, which is evidently an isomorphism. Thus if $X \times -$ reflects isomorphisms, we have $S \times \operatorname{supp} X = S$ or $S \leq \operatorname{supp} X$ for all $S \in \operatorname{Supp} \underline{X}$. Since every object maps to some $S \in \operatorname{Supp} \underline{X}$, every object has a map, necessarily unique to supp X, which means that it is terminal. On the other hand, suppose supp X is the terminal object, which we will denote 1, and suppose that $Y \xrightarrow{f} Y'$ is any map with $X \times Y \xrightarrow{X \times f} X \times Y'$ an isomorphism. We first show that f must be $\longrightarrow\!\!\!\!\!\rightarrow$.

The diagram

$$\begin{array}{ccc} X \times Y' & \longrightarrow\!\!\!\!\!\rightarrow & Y' \\ \downarrow & & \downarrow \\ X & \longrightarrow\!\!\!\!\!\rightarrow & 1 \end{array}$$

is a pullback, whence $X \times Y' \longrightarrow\!\!\!\!\!\rightarrow Y'$, which together with the commutative diagram

$$\begin{array}{ccc} X \times Y & \xrightarrow{\sim} & X \times Y' \\ \downarrow & & \downarrow\!\!\downarrow \\ Y & \longrightarrow & Y' \end{array}$$

and I. (2.5) implies that $Y \longrightarrow\!\!\!\!\!\rightarrow Y'$.

Now form

$$Y''' \xrightarrow{d} Y'' \overset{d^0}{\underset{d^1}{\rightrightarrows}} Y \xrightarrow{f} Y'$$

in which $Y'' \overset{d^0}{\underset{d^1}{\rightrightarrows}} Y$ is the kernel pair of f and $Y''' \xrightarrow{d} Y''$ is their equalizer. Exactly as in the proof of I (2.16), $X \times -$ preserves kernel pairs and equalizers, and so

$$X \times Y''' \longrightarrow X \times Y'' \rightrightarrows X \times Y \longrightarrow X \times Y'$$

is a sequence of the same type But now $X \times f \xrightarrow{\sim} \Rightarrow X \times d^o = X \times d^1$ implies that $X \times d$ is $\xrightarrow{\sim}$. By the above, this implies that d is $\twoheadrightarrow$, which implies $d^o = d^1$ and then that f is $\rightarrowtail$. By the uniqueness of the factorization, only an isomorphism can be both.

(2.12) <u>Definition</u>. Let $\underline{X}$ be a regular category with a terminal object 1. An object $X \in \underline{X}$ is said to have full support or to be fully supported if $X \twoheadrightarrow 1$. $\underline{X}$ is called fully supported if every object of $\underline{X}$ is. This is equivalent to the existence of only one partial terminal object, since the existence of a terminal object is enough to show that supports exist.

(2.13) It is clear from the results of this section that the functor Supp is a fibration, that the fibres are fully supported regular categories (and exact if the total category is), and that the transition functors are exact. This last follows from the fact that the transition functor from the fibre over S for $S \leqslant S'$ is given by $S \times -$. This functor preserves all projective limits, since $S^n = S$ for all cardinals n. Conversely, any partially ordered $\underline{P}$ together with a functor $\underline{P}^{op}$ to the category of regular (resp. exact) categories and exact functors can be pasted together to make a regular (resp. exact) category.

(2.14): <u>Proposition</u>. Every map in $\underline{X}$ may be factored f = g.h where supp h is an identity and f is a cartesian map in the fibration.

Proof. This is the essence of a fibration. Given $f: X \longrightarrow Y$, we factor it as $X \longrightarrow \text{supp } X \times Y \longrightarrow Y$. The existence of f implies $\text{supp } X \leqslant \text{supp } Y$, so $\text{supp}(\text{supp } X \times Y) = \text{supp } X$. The second factor is exactly a cartesian map.

(2.15) <u>Proposition</u>. Let $\underline{S}$ be a full subcategory of supp $\underline{X}$. Then the full subcategory of $\underline{X}$ consisting of those objects whose support lies in $\underline{S}$ is regular (and exact when $\underline{X}$ is).

Proof. Trivial.

3. Diagrams

(3.1) Let $\underline{I}$ be an (index) category and D: $\underline{I} \longrightarrow \underline{X}$ be a functor. Then we will often say that the functor D, or for emphasis, the pair ($\underline{I}$,D), is a diagram in $\underline{X}$.

(3.2) If ($\underline{I}$,D) is a diagram in $\underline{X}$ and X is an object, let (D,X) denote the set colim(Di,X), the colimit being taken over i $\in$ $\underline{I}$. Then an element of (D,X) is represented by an object i $\in$ $\underline{I}$ together with a map f: Di $\longrightarrow$ X. We may denote this (i,f) and its class by $\Vert i,f \Vert$. Then $\Vert i,f \Vert = \Vert j,g \Vert$ if f: Di $\longrightarrow$ X and g: Dj $\longrightarrow$ X are the same in the colimit. In the special case when $\underline{I}$ is filtered (the only type of diagram we will have - in fact they will all be directed sets), this means that there is a k $\in$ $\underline{I}$ and α: k $\longrightarrow$ i, β: k $\longrightarrow$ j in $\underline{I}$ such that

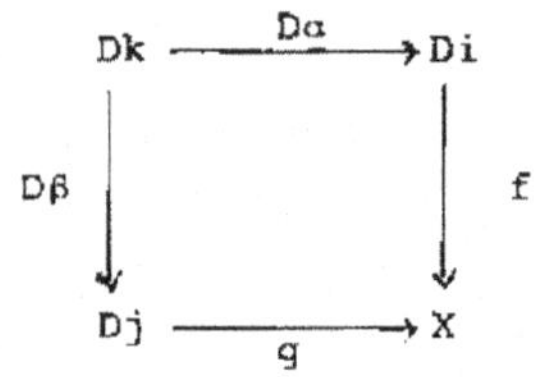

commutes. When $\underline{I}$ is not filtered, take the equivalence relation generated by that relation.

(3.3) More generally, if ($\underline{I}$,D) and ($\underline{J}$,E) are diagrams, we define (D,E) as lim(D,Ej), the limit taken over j $\in$ $\underline{J}$. In effect, an element of (D,E) is represented by choosing for each j $\in$ $\underline{J}$ a σj $\in$ $\underline{I}$ and a map fj: Dj $\longrightarrow$ Ej such that for α: $j_1 \longrightarrow j_2$ in $\underline{J}$, $\Vert \sigma j_1, E\alpha . fj_1 \Vert$ = $= \Vert \sigma j_2, fj_2 \Vert$ in (D,Ej_2). Then two families (σ,{fj}) and (τ,{gj}) represent the same element of (D,E) if for each j $\in$ $\underline{J}$, $\Vert \sigma j, fj \Vert$ = $= \Vert \tau j, gj \Vert$ as maps of D $\longrightarrow$ Ej. The composition of two such families is obvious and gives a category. Diag $\underline{X}$, of diagrams in $\underline{X}$.

(3.4) Proposition. If ($\underline{I}$,D) and ($\underline{J}$,E) are two diagrams, then (D,E) = $= \lim_{j \in \underline{J}} \operatorname{colim}_{i \in \underline{I}} (Di, Ej)$

Proof. This is just a shorthand form of the above discussion.

(3.5) If $X \in \underline{X}$, we let X also denote the diagram ($\underline{I}$,D) where I has exactly one object i and one map and Di = X. Then this embedding is obviously full and faithful. In fact, it can be easily seen that Diag $\underline{X}$ is just $(\underline{X},\underline{S})^{op}$ and that this embedding is the Yoneda embedding. However, this fact is not needed here, as we will work directly with diagrams. On account of this, we will call such a diagram either representable or the diagram represented by X.

(3.6) From now on, all diagrams will be over partially ordered sets, in fact, over inverse directed sets. In terms of functor categories, this means that we are restricting our attention to the category of finite-limit-preserving functors. If, for $i,j \in \underline{I}$ there is a map $j \longrightarrow i$, i.e. if $j \leq i$, we use (i,j) to denote it; and then, of course, $D(i,j)$: $Dj \longrightarrow Di$ is the corresponding map in the diagram.

(3.7) Recall that every $f: X \longrightarrow Y$ can be factored in the form

$$X \xrightarrow{h} X \times \operatorname{supp} X \xrightarrow{g} Y.$$

We will say that f is special if h is $\longrightarrow\!\!\!\!\rightarrow$.

(3.8) Proposition: Special morphisms are stable under composition and pullbacks.

Proof. Let $X \longrightarrow Y$ and $Y \longrightarrow Z$ be special. Then $X \twoheadrightarrow \operatorname{supp} X \times Y$ and $Y \twoheadrightarrow \operatorname{supp} Y \times Z$ give $\operatorname{supp} X \times Y \twoheadrightarrow \operatorname{supp} X \times \operatorname{supp} Y \times Z = \operatorname{supp} X \times Z$. This, together with I. (2.8), gives the first result. As for the second, if $X \longrightarrow Y$ is special and we form a pullback

$$\begin{array}{ccccc} X & \twoheadrightarrow & \operatorname{supp} X \times Y \times_Y Y' & \longrightarrow & Y' \\ \downarrow & & \downarrow & & \downarrow \\ X & \twoheadrightarrow & \operatorname{supp} X \times Y & \longrightarrow & Y \end{array}$$

then $\operatorname{supp} X \times Y \times_Y Y' \cong \operatorname{supp} X \times Y'$.

(3.9) Given a diagram $(\underline{I},D)$, we define a new diagram $(\underline{I}_S, S_S)$ for any $S \in \operatorname{Supp} \underline{X}$ by letting $\underline{I}_S = \{i \mid \operatorname{supp} Di \geq S\}$ and $D_S i = Di \times S$. We see that D_S can be thought of as being a functor $\underline{I}_S \longrightarrow \underline{X}_S$, where the latter denotes the full subcategory of all objects whose support is S.

(3.10) Given a diagram $(\underline{I},D)$ we say it is P-diagram if it satisfies:

P1) $\underline{I}_S$ is an inf semilattice for all $S \in \operatorname{Supp} \underline{X}$.

P2) For any $i \in \underline{I}$ and any special morphism $f: X \longrightarrow Di$, there is a $j \leq i$ with $D(i,j) = f$ (and of course $Dj = X$).

The diagram $(\underline{I},D)$ is called an A-diagram if it satisfies:

A1) = P1).

A2) For any $i < j$, the interval $(i,j] = \{k \mid i < k \leq j\}$ is finite.

A3) For any $i < j$, the natural map $Di \longrightarrow \lim(D \mid (i,j])$ is special.

(3.11) It should be noted that these definitions are not isomorphism invariant and should be supplemented by saying that a diagram isomorphic to one of the above type is of that type also. It would be useful to discover, purely in terms of the functors represented, what these definitions mean.

(3.12) Proposition. Let $(\underline{I},D)$ be a P-diagram (resp. A-diagram) in $\underline{X}$.

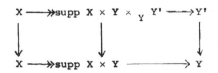

Then $(\underline{I}_S, D_S)$ is a P-diagram (resp. A-diagram) in $\underline{X}_S$.

Proof. The condition P1) = A1) is evidently designed to be inherited in this way. If $f\colon X \longrightarrow D_S i$ is special, supp X = S clearly is equivalent to $X \twoheadrightarrow D_S i$. There must exist $j < i$ with $D(i,j) = f$. We have supp Dj = S, so $j \in I_S$ and $D_S j = Dj$. Thus P2) is inherited. If $(\underline{I}, D)$ is an A-diagram, $(\underline{I}_S, D_S)$ satisfies A1 as above and A2 is clear. Then $Di \longrightarrow \lim D|(i,j]$ being special implies that

$$Di \twoheadrightarrow \text{supp } Di \times \lim D|(i,j],$$

and if supp $Di \geq S$,

$$\begin{aligned} S \times Di &\twoheadrightarrow S \times \text{supp } Di \times \lim D|(i,j] \\ &= S \times \lim D|(i,j] \\ &= \lim D_S|(i,j], \end{aligned}$$

since supp $Dk \geq S$ for all $k > i$ and $S \times -$ is an exact functor.

(3.13) Proposition. Let $(\underline{I}, D)$ be an A-diagram. Then $D(j,i)$ is special for $i < j$. Also $D_S(j,i)$ is $\twoheadrightarrow$ for all $i < j$ such that supp $Di \geq S$.

Proof. Since the interval $(i,j]$ is finite, there is a finite chain $i = i_o < i_1 < \dots < i_n = j$ such that each $(i_r, i_{r+1}]$ has only one element, namely i_{r+1}, and then A3 implies that $Di_r \longrightarrow Di_{r+1}$ is special. Then $D(j,i)$, being the composite of these, is special also. The last statement is obvious, since a special morphism between two objects of the same support is $\twoheadrightarrow$.

(3.14) Proposition. Let $(\underline{I}, D)$ be a P-diagram. Then for any $S \in \text{Supp } \underline{X}$,

$$(D_S, -)\colon \underline{X} \longrightarrow \underline{S}$$

is exact.

Proof. Since $\underline{I}_S$ is inverse directed, it evidently preserves finite limits. If $f\colon X \twoheadrightarrow Y$, then supp X = supp Y. Let $\|i,g\|\colon D_S \longrightarrow Y$ be a map. Since the pullback of

$$\begin{array}{ccc} & & X \\ & & \Big\downarrow \\ D_S i & \xrightarrow{g} & Y \end{array}$$

comes equipped with a $\longrightarrow\!\!\!\!\!\rightarrow D_S i$, it is represented in the diagram, so there is a commutative diagram

$$\begin{array}{ccc} D_S j & \xrightarrow{h} & X \\ {\scriptstyle D_S(i,j)} \Big\downarrow & & \Big\downarrow \\ D_S i & \xrightarrow{g} & Y. \end{array}$$

Then $\|j,h\|: D_S \longrightarrow X$ is a map such that $(D_S,f)\|j,h\| = \|j,g.D_S(i,j)\| = \|i,g\|$, which implies that (D_S,f) is onto.

(3.15) Proposition. Let $(\underline{I},D)$ be a P-diagram. For each $i \in \underline{I},S$, $\|i, D_S i\|: D_S \longrightarrow D_S i$ is an epimorphism.

Proof. As pointed out in (3.13), every map in the diagram D_S is $\longrightarrow\!\!\!\!\!\rightarrow$. If $f,g: Di \longrightarrow X$ are distinct, then for all $j < i$, $D(i,j)f \neq D(i,j).g$. Evidently every diagram is the limit of representable diagrams and an inverse limit of monomorphisms is a monomorphism.

4. The Lubkin completion process.

(4.1) In this section we show how to "complete" a given diagram to a P-diagram. This construction was first described by Lubkin in his original proof of the abelian category imbedding, [Lu]. As a matter of fact, Lubkin observed then that there was nothing inherently abelian in his proof. Lubkin even stated a non-abelian embedding theorem, although based on the notion of ordinary, rather than regular, epimorphisms.

(4.2) Let $(\underline{I},D)$ be a diagram, $i_o \in \underline{I}$ and $f\colon X \longrightarrow Di_o$ be a map in $\underline{X}$. We describe a new diagram $\mathrm{Lub}(\underline{I},D,i_o,f) = (\underline{I}',D')$ as follows. Let $\underline{I}^*$ be a partially ordered set disjoint from and order isomorphic to $\{i \in \underline{I} \mid i \leqslant i_o\}$, by a map $i \longleftrightarrow i^*$. Let $\underline{I}'$ denote $\underline{I} \cup \underline{I}^*$, in which each component has its own order and moreover $i^* < j$ if and only if $i \leqslant j$. In particular, $i^* < i$, and the order is generated by that relation together with the orders in $\underline{I}$ and $\underline{I}^*$. We define D' by $D'|\underline{I} = D$, $D'i_o^* = X$, $D'(i_o,i_o^*) = f$, and for $i \leqslant i_o$, $D'i^*$ is defined so that the diagram

$$\begin{array}{ccc} D'i^* & \xrightarrow{\;D'(i_o^*,i^*)\;} & X = D'i \\ {\scriptstyle D'(i,i^*)}\Big\downarrow & & \Big\downarrow{\scriptstyle f} \\ Di & \xrightarrow[\;D(i_o,i)\;]{} & Di_o \end{array}$$

is a pullback. D' is defined on maps $i^* \longrightarrow i_o^*$ and $i^* \longrightarrow i$ as shown. For $i \leqslant j \leqslant i_o$, $D'(j^*,i^*)$ is uniquely induced by a pullback and $D'(j,i^*)$ is defined as $D'(j,j^*).D'(j^*,i^*) = D(j,i).D'(i,i^*)$. This last equality is a consequence of the definition of $D'(j^*,i^*)$ as a map into a pullback.

(4.3) Let $(\underline{I},D)$ and $(\underline{I}',D')$ be diagrams. We say that $(\underline{I}',D')$ is a Lubkin-extension of $(\underline{I},D)$ if there is some $i_o \in \underline{I}$ and $f\colon X \longrightarrow Di$

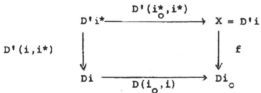

with $(\underline{I}', D') = \mathrm{Lub}(\underline{I}, D, i_o, f)$. In particular, this means that $\underline{I} \subset \underline{I}'$ and $D'|\underline{I} = D$.

(4.4) Let n be an ordinal number. A sequence $\{(\underline{I}_m, D_m) \mid m \leq n\}$ of diagrams is called a Lubkin-sequence if for each m, $(\underline{I}_{m+1}, D_{m+1})$ is a Lubkin-extension of $(\underline{I}_m, D_m)$ and if for each limit ordinal m, $\underline{I}_m = \bigcup_{p<m} \underline{I}_p$; $D_m|\underline{I}_p = D_p$.

(4.5) Let $(\underline{I}, D)$ be a diagram. If n is an ordinal number and $\{f_m \mid m < n\}$ is a sequence of morphisms $f_m: X_m \longrightarrow Di_m$, we define a Lubkin-sequence by letting $(\underline{I}_o, D_o) = (\underline{I}, D)$, and for each m, $(\underline{I}_{m+1}, D_{m+1}) = \mathrm{Lub}(\underline{I}_m, D_m, i_m, f_m)$, while for each limit ordinal m, $\underline{I}_m = \bigcup_{p<m} \underline{I}_p$, $D_m|\underline{I}_p = D_p$.

(4.6) Let $(\underline{I}, D)$ be a diagram. Let n_1 be an ordinal such that there is a 1-1 correspondence $m \longmapsto f_m$ between the ordinals $m < n$ and the set of all special morphisms whose codomain is a Di for $i \in \underline{I}$. Then applying the above construction, we get a diagram $(\underline{I}_{n_1}, D_{n_1})$. This diagram has the property that given $i \in \underline{I}$ and $f: X \longrightarrow Di$ special, there is some $j \in \underline{I}_{n_1}$ such that $j < i$ and $f: X \longrightarrow Di = D_{n_1}(i,j): D_{n_1} j \longrightarrow Di$. Now let n_2 be an ordinal such that there is a 1-1 correspondence $m \longmapsto f_m$ between all the ordinals $n_1 \leq m < n_2$ and the set of all special morphisms whose domain is a $D_{n_1} i, i \in \underline{I}_{n_1}$. Extend the Lubkin-sequence $\{(\underline{I}_m, D_m) \mid m \leq n_1\}$ to one defined for $m \leq n_2$ by applying the process of (4.5) beginning with $(\underline{I}_{n_1}, D_{n_1})$. Then we may continue in this way with ordinals $n_2, n_3, \ldots$. Let $n = \sup\{n_i \mid i \in \omega\}$. By letting $\underline{I}_n = \bigcup_{m<n} \underline{I}_m$, $D_n|\underline{I}_m = D_m$, we construct a Lubkin sequence $\{(\underline{I}_m, D_m) \mid m \leq n\}$ with the property that for all special $f: X \longrightarrow Di$, $i \in \underline{I}_n$, there is a $j < i$ in $\underline{I}_n$ such that $f: X \longrightarrow Di = D(i,j): Dj \longrightarrow Di$.

The diagram $(\underline{I}_n, D_n)$ will be called a Lubkin completion of $(\underline{I}, D)$.

(4.7) Proposition. Let $(\underline{I}, D)$ be a diagram in which $\underline{I}_S$ is an inf semilattice for each $S \in \text{Supp } \underline{X}$. Then a Lubkin completion of it is a P-diagram.

Proof. P1) is an inductive property, so it suffices to consider a single Lubkin extension. Let $(\underline{I}, D)$ satisfy P1) and $(\underline{I}', D') =$ $= \text{Lub}(\underline{I}, D, i_o, f)$. Let $i \wedge j$ denote the inf of two elements of $\underline{I}_S$. If $i < i_o$ and $i \in \underline{I}_S$, then $i_o \in \underline{I}_S$ also and supp $Di^* =$ supp $Di \cap$ supp X, where X is the domain of f. If supp X is not $\geq S$, then $\underline{I}'_S = \underline{I}_S$. If supp $X \geq S$, then supp $Di^* \geq S$ if and only if supp $Di \geq S$. Now if $i,j \in \underline{I}_S$, $i \wedge j \in \underline{I}'_S$, being the same as in $\underline{I}_S$. If $i,j \in \underline{I}_S$, $i \leq i_o$, $i^* \wedge j = (i \wedge j)^*$ and is in $\underline{I}_S$ when i^*_o is. If also $j \leq i_o$, $i^* \wedge j^* =$ $(i \wedge j)^*$ as well. As for P2), this is what the Lubkin completion is all about. Supposing that $i \in \underline{I}_n$ and $f: X \longrightarrow Di$ is special, then $i \in \underline{I}_{n_r}$ for some $r \in \omega$ and $f = f_m$ for some ordinal m such that $n_r < m < n_{r+1}$. Then f is represented in the diagram $(\underline{I}_m, D_m)$ and thereafter.

(4.8) Proposition. Suppose $(\underline{I}, D)$ is an A-diagram. Then any Lubkin extension of it is an A-diagram.

Proof. Let $(\underline{I}', D') = \text{Lub}(\underline{I}, D, i_o, f)$. We have just seen that A1) = P1) is preserved by Lubkin extension. As for A2), if $i,j \in \underline{I}$, $(i,j]$ is the same in $\underline{I}$ and $\underline{I}'$. If $i,j \in \underline{I}$, $i < i_o$, $(i^*,j] = (i^*,(j \wedge i)^*] \cup [i,j]$ and the first term is order isomorphic to $(i, j \wedge i]$. If $j < i_o$ also, $(i^*,j^*]$ is order isomorphic to $(i,j]$. To show A3) is satisfied, we consider the cases.

Case 1. $i < j$ in $\underline{I}$. This follows directly from the fact that $D'|\underline{I} = D$.

Case 2. $i^* < j^*$. This case is a simple application of the fact that limits commute with limits to show that

$$\begin{array}{ccc} D'i^* & \longrightarrow & \lim D'|(i^*,j^*] \\ \downarrow & & \downarrow \\ Di & \longrightarrow & \lim D|(i,j] \end{array}$$

is a pullback. Then since the bottom arrow is special, so is the top.

Case 3. $i^* < j$ but $i = i_o \wedge j$. In this case, $(i^*,j] = [i,j]$ and so $\lim D'|(i^*,j] = Di$. Then since f is special, so is $D'i^* \longrightarrow Di$.

Case 4. $i^* < j$ and $i < i_o \wedge j$. I claim that in this case Di^* is the limit under consideration. To see this let $j_o = j \wedge i_o$, and suppose we are given $g(k): Y \longrightarrow Dk$ for each $k \in [i,j]$ and $g(k^*)$: $Y \longrightarrow Dk^*$ for each $k \in (i^*,j_o^*]$, which constitute a coherent family. Then $D'(j_o,j_o^*).g(j_o^*) = D'(j_o,i).g(i)$, so that since

$$\begin{array}{ccc} D'i^* & \longrightarrow & D'j_o^* \\ \downarrow & & \downarrow \\ Di & \longrightarrow & Dj_o \end{array}$$

is a pullback, there is a unique $g: Y \longrightarrow D'i^*$ such that $D'(i,i^*).g =$ $= g(i)$ and $D'(j_o^*,i^*).g = g(j_o^*)$. If $k \in [i,j]$, then $g(k) = D(k,i).g(i)$, so that $D'(k,i^*).g = D(k,i).D'(i,i^*).g = D(k,i).g(i) = g(k)$. If $k^* \in (i,j_o^*]$, then to show that $D'(k^*,i^*).g = g(k^*)$, we use the fact that

$$\begin{array}{ccc} D'k^* & \longrightarrow & D'j_o^* \\ \downarrow & & \downarrow \\ Dk & \longrightarrow & Dj_o \end{array}$$

is a pullback. We have $D'(j_o^*,k^*).D'(k^*,i^*).g = D'(j_o^*,i^*).g = g(j_o^*) =$ $= D'(j_o^*,k^*).g(k^*)$ and $D'(k,k^*).D'(k^*,i^*).g = D'(k,i^*).g =$

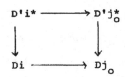

$= D(k,i).D'(i,i^*).g = D(k,i).g(i) = g(k) = D'(k,k^*).g(k^*)$.

(4.9) Corollary. A Lubkin completion of an A-diagram is simultaneously an A- and P-diagram.

5. The embedding.

(5.1) We are now ready to describe the embedding. The functor $\underline{X}(1,-)$ is represented by the diagram $D_o\colon \underline{I}_o \longrightarrow \underline{X}$ in which $\underline{I}_o$ has one object and D_o at that object is the terminal object 1. This is evidently an A-diagram and we let $(\underline{I},D)$ be a Lubkin completion of it. We let $\underline{C}$ be the category whose objects are the non-empty subobjects of 1, and whose morphisms are defined by

$$\underline{C}(S_1,S_2) = (D_{S_1},D_{S_2}):$$

that is, morphisms (as defined in (3.3)) between the diagrams $(\underline{I}_{S_1},D_{S_1})$ and $(\underline{I}_{S_2},D_{S_2})$. This is equivalent to natural transformations between the functors represented by the diagrams. Composition in $\underline{C}$ is just the composition of natural transformations. Note that $\underline{C}(S_1,S_2) = \emptyset$ unless $S_1 \leq S_2$, which means that there is a functor $\underline{C} \longrightarrow \text{Supp } \underline{X}$. We define $U\colon \underline{X} \longrightarrow (\underline{C}^{op},\underline{S})$ by $(UX)S = (D_S,X)$, the mapping described in (3.2). Composition of natural transformations (recall that this is really natural transformations between $(X,-)$ and $(D_S,-)$) makes this functorial in $\underline{X}$ and (contravariantly) in $\underline{C}$. Since limits and colimits in functor categories are computed elementwise, it follows that U is exact as long as $(U-)S$ is for each S. That functor is $(D_S,-)$.

(5.2) Proposition. U is exact.

Proof. See (3.14).

(5.3) Proposition. Let $E\colon \underline{J} \longrightarrow \underline{X}_S$ be a P-diagram and $F\colon \underline{K} \longrightarrow \underline{X}_S$ be an A-diagram. Let $k_o \in K$ and

$$E \xrightarrow{\|j_o,f\|} Fk_o$$

be a map. Then it extends to a map $E \longrightarrow F$. This means that there is a map $E \longrightarrow F$ such that

$$E \longrightarrow F$$

$$\|j_o,Ej_o\| \downarrow \qquad \qquad \downarrow \|k_o,Fk_o\|$$

$$Ej_o \longrightarrow Fk_o$$

commutes, since always $f.\|j_o,Ej_o\| = \|j_o.f\|$.

Note that we use the name of an object to denote also its identity map.

Proof. First we observe that F (like any diagram based on an inverse directed set) is isomorphic to the diagram gotten by truncating F above k_o: That is, replacing $\underline{K}$ by $\{k \mid k \geqslant k_o\}$ and restricting F. This new diagram, moreover, satisfies the conditions for being an A-diagram itself (not merely being isomorphic to one). Thus we may suppose that k_o is terminal in $\underline{K}$. Next we observe that $E = E_S$ represents an exact functor of $\underline{X} \longrightarrow \underline{S}$. This means that the $\underline{S}$ diagram $(\underline{K},\widetilde{F})$ defined by $\widetilde{F}k = (E,Fk)$ is an A-diagram in $\underline{S}$, since exact functors preserve the properties defining an A-diagram, finite limits as well as regular epimorphisms (which are what special maps reduce to in $\underline{X}_S$). Since $(E,F) = \lim(E,Fk)$, then $(E,F) = \lim \widetilde{F}k$, taken over $k \in \underline{K}$. Hence this proposition is reduced to the following special case (when $E = 1$ and $\underline{X} = \underline{S}$).

(5.4) **Proposition.** Let $(\underline{K},F)$ be an A-diagram in $\underline{S}$ and $k_o \in \underline{K}$ be terminal. Then $\lim F \longrightarrow Fk_o$ is onto.

Proof. We choose a point of Fk_o which we will denote by $p(k_o)$. We consider families $(\underline{L},p(\underline{L}))$ in which $\underline{L}$ is a full subset of $\underline{K}$ that is, a subset with the restricted order) and $p(\underline{L}) = \{p(l) \mid l \in \underline{L}\}$ is a point of $\lim F/\underline{L}$ subject to the following conditions.

a) $k_o \in \underline{L}$.

b) $p(k_o)$ is the already given point.

c) For $k \in \underline{K}$, $l \in \underline{L}$, $l < k \Longrightarrow k \in \underline{L}$.

This family is partially ordered in the obvious way: $(\underline{L}_1, p(\underline{L}_1)) < (\underline{L}_2, p(\underline{L}_2))$ if $\underline{L}_1 \subset \underline{L}_2$ and $p(\underline{L}_2)|\underline{L}_1 = p(\underline{L}_1)$. This set is inductive; the only thing non-trivial is showing that a union of a nested family has a point of the limit. But the test of whether a point of $\{F\ell \mid \ell \in \underline{L}\}$ is a point of the inverse limit involves only two indices at a time, and in an inductive union the satisfaction of such a test is inherited. Hence there is a maximal $(\underline{L}, p(\underline{L}))$ among the family. We need only show that $\underline{K} = \underline{L}$. If not, there is $k \in \underline{K}$, $k \notin \underline{L}$: Since the interval $(k, k_o]$ is finite and $k_o \in \underline{L}$, there must be some $k \notin \underline{L}$ for which $(k, k_o] \subset \underline{L}$. But since

$$Fk \longrightarrow \lim F|(k, k_o]$$

is onto and $\{p(\ell) \mid \ell \in (k, k_o]\}$ is an element of that inverse limit, there is a $p(k) \in Fk$ such that for all $k' \in (k, k_o]$, i.e. all $k' > k$, $F(k', k)p(k) = p(k')$. By condition c) above, no element of $\underline{L}$ precedes k, so that in fact $p(\underline{L}) \cup \{p(k)\}$ is a point of $\lim F|\underline{L} \cup \{k\}$. Clearly the conditions a), b), and c) above are satisfied and we have constructed a proper extension of $(\underline{L}, p(\underline{L}))$, which is a contradiction.

(5.5) Now for an object $X \in \underline{X}$ with support S. Let $(\underline{I}, D)$ be the diagram constructed in (5.1). Since $X \longrightarrow 1$ factors as $X \twoheadrightarrow S \rightarrowtail 1$, there is some $i_o \in \underline{I}$ with $Di_o = X$. Let $\underline{J} = \{i \in \underline{I}_S \mid i \leq i_o\}$. Let $E = D|\underline{J}$. Evidently $(\underline{J}, E) \cong (\underline{I}_S, D_S)$, and $(\underline{J}, E)$ is easily seen to be both an A- and a P-diagram. Let $F: \underline{J} \longrightarrow \underline{X}$ be the functor whose value at $i \in \underline{J}$ is the kernel pair of $E(i_o, i) = D(i_o, i)$. Since Di and Di_o have the same support, this amounts to saying that

$$Fi \underset{d^1 i}{\overset{d^o i}{\rightrightarrows}} Ei \xrightarrow{E(i_o, i)} Ei_o = X$$

is exact.

(5.6) __Proposition.__ The diagram $(\underline{J}, F)$ is an A-diagram.

Proof. A1) and A2) are obvious. Let $k < j \in \underline{J}$. Since limits commute with limits,

$$\lim F|(k,j] = \lim(E \times_X E)|(j,k] = \lim E|(j,k] \times_X \lim E|(j,k].$$

Since $Ej \twoheadrightarrow \lim E|(j,k]$, the result $Ej \times_X Ej \twoheadrightarrow \lim(E \times_X E)|(j,k]$ follows from I.(2.2).

(5.7) Proposition. The diagram

$$F \underset{d^1}{\overset{d^0}{\rightrightarrows}} E \xrightarrow{\|i_0,X\|} X$$

is a coequalizer.

Proof. Since every diagram is a limit of objects of $\underline{X}$, it is sufficient to show this for maps into them. Suppose $\|j,g\|: E \longrightarrow Y$ is a map coequalizing d^0 and d^1. This means that $\|j,g.d^0\| = \|j,g.d^1\|$, and since $F \xrightarrow{\|j,Fj\|} Fj$ is an epimorphism (see (3.15)), it follows that $g.d^0 = g.d^1$. But

$$\begin{array}{ccc} Fj \underset{d^1 j}{\overset{d^0 j}{\rightrightarrows}} & Ej & \xrightarrow{E(i_0,j)} X \\ & \downarrow g & \\ & Y & \end{array}$$

is a coequalizer and hence there is induced $f: X \longrightarrow Y$ with $f.E(i_0,j) = g$. Since $E(i_0,j)$ is a map in the diagram, it represents the map $\|i_0,X\|: E \longrightarrow X$. Uniqueness of f follows from (3.15).

(5.8) Proposition. Let $G: \underline{K} \longrightarrow \underline{X}$ be any diagram and F the diagram constructed in (5.5). Given two distinct maps $F \rightrightarrows G$, there is a map $E \longrightarrow F$ with $E \longrightarrow F \rightrightarrows G$ also distinct.

Proof. It is sufficient, as above, to consider the case when G is an object of $\underline{X}$, say $G = Y$. Let the two maps be $\|i,f\|: F \longrightarrow Y$ and $\|j,g\|: F \longrightarrow Y$. By choosing $k \geqslant i,j$ we may suppose that $i = j$.

Since $Fi \xrightarrow{E(i_o,i).d^o i} \!\!\!\!\rightarrow X$, there is some $\mathcal{X} \in \underline{J}$ such that $E\mathcal{X} = Fi$. Since F is an A-diagram (see (5.6)), the map $E\mathcal{X} \longrightarrow Fi$ can be extended to a map $E \longrightarrow F$, giving a commutative diagram

$$\begin{array}{ccc} E & \longrightarrow & F \\ \downarrow & & \downarrow \\ E\mathcal{X} & \xrightarrow{=} & Fi \end{array}$$

and $E \longrightarrow E\mathcal{X}$ an epimorphism. Since $Fi \rightrightarrows Y$ are distinct, so are $E \longrightarrow E\mathcal{X} \rightrightarrows Fi \longrightarrow Y$, and then $E \longrightarrow F \rightrightarrows Y$.

(5.9) <u>Proposition</u>. U is full and faithfull.

Proof. Suppose $X \underset{g}{\overset{f}{\rightrightarrows}} Y$ and $Uf = Ug$. If $Z \xrightarrow{e} X$ is the equalizer, this implies that Ue is an isomorphism. If $S = \text{supp } X$, $(UZ)S \cong (UX)S$ and $(UX)S \neq \emptyset$ implies that $(UZ)S \neq \emptyset$ and that $S \leq \text{supp } Z$, while clearly $\text{supp } Z \leq S$. Now choose a vertex $i \in \underline{I}_S$ with $Di = X$. By the isomorphism, the element $\| i,X \| \in (UX)S$ must come from $(UZ)S$ and be represented by some $\| j,h \|$. By choosing $k = i \wedge j$ and observing that $D_S \longrightarrow D_S k$ is epi (see (3.15)), we have a commutative diagram

$$\begin{array}{ccc} Dk & \xrightarrow{h.D(j,k)} & Z \\ {\scriptstyle D(i,k)}\downarrow & & \downarrow{\scriptstyle e} \\ Di & \xrightarrow{=} & X \end{array}$$

from which we see that e is $\twoheadrightarrow$. Since e is also an equalizer, this implies that e is an $\xrightarrow{\sim}$ and that $f = g$.

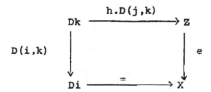

Now suppose that $\varphi\colon UX \longrightarrow UY$ is a natural transformation of functors. Taking $S = \text{supp } X$, we see that $\varphi S\colon (UX)S \longrightarrow (UY)S$, and since $(UX)S \neq \emptyset$, $(UY)S \neq \emptyset$ and $S \leq \text{supp } Y$. If $s\colon X \twoheadrightarrow S$ is the map (there is only one), then $(\varphi, Us): UX \longrightarrow UY \times US = U(Y \times S)$ is also natural. If we show that $(\varphi, Us) = U(f,s), f\colon X \longrightarrow Y$, then $(\varphi, Us) = (Uf, Us)\colon UX \longrightarrow UY \times Us$ and $\varphi = p_2.(\varphi, Us) = p_1.(Uf, Us) = Uf$. Hence it is sufficient to consider the case that supp $Y = S$ as well. Let $(\underline{J}, E)$ and $(\underline{J}, F)$ be the diagrams constructed in (5.5) above. Then $(UX)S = (F,X)$ and $(UY)S = (F,Y)$. Let d denote $\|i_o, X\|$: $E \longrightarrow X$. Then by (5.7),

$$F \underset{d^1}{\overset{d^o}{\rightrightarrows}} E \xrightarrow{\quad d \quad} X$$

is a coequalizer. Now the map d represents an element, also denoted d, of UX, and is transformed into an element $\varphi(d)\colon E \longrightarrow Y$. If $\varphi(d).d^o \neq \varphi(d).d^1$ as maps $F \longrightarrow Y$, there would exist, by (5.8), a map $g\colon E \longrightarrow F$ such that $\varphi(d).d^o.g \neq \varphi(d).d^1.g$. But the statement that φ is natural means that for any map $S \longrightarrow S$ in $\underline{C}$, that is to say, any natural transformation $u\colon E \longrightarrow E$, and for any $h\colon E \longrightarrow X$, $\varphi(h.u) = \varphi(h).u$. But $d^o.g$ and $d^1.g$ are maps $E \longrightarrow E$, and so we have $\varphi(d).d^o.g = \varphi(d.d^o.g) = \varphi(d.d^1.g) = \varphi(d).d^1.g$, which is a contradiction. Thus $\varphi(d).d^o = \varphi(d).d^1$, and by the property of equalizers, there is induced a map $f\colon X \longrightarrow Y$ with $f.d = \varphi(d)$. Now suppose $e\colon E \longrightarrow X$ represents some other element of $(UX)S$. Since E is an A- and P-diagram, $e\colon E \longrightarrow X$ can be extended to $v\colon E \longrightarrow E$ such that $d.v = e$. Then $\varphi(e) = \varphi(d.v) = \varphi(d).v = f.d.v = f.e$. Hence $\varphi = Uf$. This completes the proof.

(5.10) <u>Proposition</u>. For each object X of $\underline{X}$, UX is a regular quotient of a representable functor.

Proof. Let $S = \text{supp } X$. Choose an index $i \in \underline{I}_S$ with $D_S i = X$ and let $d = \|i, X\|\colon D_S \longrightarrow X$. By (5.3), we have for any P-diagram E,

$(E,D_S) \twoheadrightarrow (E,X)$. In particular, this holds for $E = D_{S'}$, and so

$$(D_{S'},D_S) \twoheadrightarrow (D_{S'},X),$$

or

$$\underline{C}(S',S) \twoheadrightarrow (UX)S',$$

which means that $\underline{C}(S',-)$ maps onto UX, or that UX is a regular quotient of $\underline{C}(S',-)$.

With this we have completed the proof of (1.6) as well as of all the other results stated in section 1.

(5.11) Remark. It seems worthwhile to make two additional remarks about this embedding. First, as a colimit of a directed set of representable functors, it does more than merely preserve the finite limits that exist. Rather it will preserve the finite limits in any reasonable finite limit completion of the category, e.g. that described in I.(4.5). The second is that as a consequence of the fact that $D_S \twoheadrightarrow D_S i$ for each i, the functor commutes with intersections of any family of subobjects of an object which have an intersection. This property is apparently a completely accidental consequence of the construction and it is not known what, if any, use it might have.

(5.12) If $\underline{V}$ is an exact closed category with exact direct limits and a faithful underlying functor, then by interpreting the $\underline{S}$ valued functor as taking values in $\underline{V}$, we get a $\underline{V}$-valued exact (not full) embedding which reflects isomorphisms. If $\underline{V}$ is the form $\underline{S}^{\mathbb{T}}$, where $\mathbb{T}$ is a commutative triple of finite rank, this is satisfied and one may even see directly that the full embedding lifts to a full exact embedding into a $\underline{V}$-valued functor category.

6. Diagram chasing.

(6.1) When one has an embedding theorem of this sort, the obvious thing to do with it is to chase diagrams. In the abelian cases this was usually cited as one of the main applications. In fact, however, in the abelian case, most of the diagrams can be chased almost as easily in the original abelian category. In fact most of the diagrams to be chased seem to involve, one way or another, the snake lemma. (I am loosely using the term "diagram-chasing" to include "diagram filling" as well.) As seen in the next two chapters, the non-abelian case offers diagrams of both greater variety and greater difficulty. This seems to be largely because exact sequences involve kernel pairs, rather than kernels; coequalizers, rather than cokernels.

(6.2) One further point, equally valid in the abelian and non-abelian case, should be mentioned here. The embedding theorem is valid for small (or locally presentable) regular categories. There are three possible ways around this difficulty for large categories, of which at least two work and one is set-theoretically unassailable. Taking that one first, any diagram, any set of objects, can be extended to a full regular (resp. exact) subcategory by a more - or - less evident process. Given a set of objects, make a full subcategory. Add to this this

a) the kernel pair of any map,

b) the regular image of any map (equivalent to the coequalizer of its kernel pair), and

c) the pullback of any pair of maps like

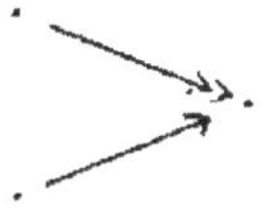

Each of the processes adds a set of objects whose number is (roughly)

the set of maps of the given subcategory.Now iterate this countably many times and take the union. The result will evidently be a full, small, regular (resp. exact) subcategory. If the original category had finite limits we could obviously modify this to give finite limits to this subcategory.

(6.3) A second possibility is to relate everything to Grothendieck universes. If a category is large in one universe, it is small in the next and can be embedded in a functor category there. Or it can first be embedded into a locally presentable category. If $\underline{S}$ is the first universe (which may as well be identified with its category of sets) and $\underline{S}^*$ is an enlargement, the embedding of $\underline{X}$ into all $\underline{S}$-continuous functors of $\underline{X}^{op} \longrightarrow \underline{S}^*$ is evidently $\underline{S}$-continuous and the functor category is locally presentable, since $\underline{X}$ is embedded as generators, each of rank $\leq$ to the cardinal of $\underline{S}$ as an object of $\underline{S}^*$.

(6.4) The final way is more speculative but would be the most satisfactory (or, anyway, the most satisfying) if it worked. It is possible that every regular category $\underline{X}$ possesses a class of exact functors $U: \underline{X} \longrightarrow \underline{S}$, $U \in \underline{U}$, with the following property. Every class $\{\varphi U \mid U \in \underline{U}\}$ of maps $UX \xrightarrow{\varphi U} UY$ for which each natural transformation $\alpha: U \longrightarrow U'$ gives a commutative diagram

$$\begin{array}{ccc} UX & \xrightarrow{\varphi U} & UX' \\ {\scriptstyle \alpha X}\downarrow & & \downarrow{\scriptstyle \alpha X'} \\ U'X & \xrightarrow{\varphi U'} & U'X' \end{array}$$

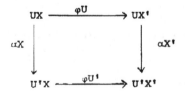

implies the existence of a unique $f: X \longrightarrow Y$ such that $\varphi U = Uf$ for all $U \in \underline{U}$. Since a class $\underline{U}$ is a collectively full and faithful family, a diagram can be chased by applying every such U. "Every" is, in this context, the same as "any" and can be supposed for purposes of verification to be just one. It is not known whether such a class $\underline{U}$ always exists.

(6.5) Whichever strategem is adopted doesn't change the fact that certain types of diagram chasing in regular categories can be carried out in functor categories. Strict diagram chasing (that is, not involving filling-in, but only commutativity) can be carried out in $\underline{S}$, since the evaluating functors $(\underline{C}^{op},\underline{S}) \longrightarrow \underline{S}$ given by evaluativy at the objects of $\underline{C}$ form a family of exact functors which are collectively faithful. In fact more is true.

(6.6) Proposition. The evaluation functors $(\underline{C}^{op},\underline{S}) \longrightarrow \underline{S}$ for $C \in \underline{C}$ collectively are faithful, exact, reflect isomorphisms and reflect equivalence relations.

Proof. That they are faithful is clear, since equality of natural transformations is defined that way. The evaluations preserve all limits and colimits (limits and colimits are calculated "pointwise"), so exactness is also clear. For similar reasons they reflect isomorphisms (collectively). Finally suppose $F \longrightarrow G \times G$ is such that FC is an equivalence relation on GC for all $C \in \underline{C}$. First, $FC \rightarrowtail (G \times G)C = GC \times GC$ implies that $F \rightarrowtail G \times G$. Next, the coequalizer $F \rightrightarrows G \longrightarrow H$ is computed pointwise so that $FC \rightrightarrows GC \longrightarrow HC$ is a coequalizer for each $C \in \underline{C}$. But the kernel pair of $GC \longrightarrow HC$ is just FC, which means that $F \rightrightarrows G$ is a kernel pair, a fortiori an equivalence relation.

(6.7) Corollary. Let $\underline{X}$ be a small (or locally presentable) regular category. Then there is a family of exact functors $U_i : \underline{X} \longrightarrow \underline{S}$, $i \in I$, which collectively are faithful, reflect isomorphisms, and reflect equivalence relations. If, in addition, $\underline{X}$ is exact, then these U_i preserve the coequalizer of any pair of maps $X \overset{d^o}{\underset{d^1}{\rightrightarrows}} Y$ such that the image of $(U_i d^o, U_i d^1): U_i X \longrightarrow U_i Y \times U_i Y$ is an equivalence relation for each $i \in I$.

Proof. If $U: \underline{X} \longrightarrow (\underline{C}^{op}, \underline{S})$ is full, faithful, and exact, we let I be the objects of $\underline{C}$ and U_i be U followed by evaluation at the corresponding object. Then every thing but the last statement is clear. To see that, suppose d^o and d^1 are as above. Then we can factor (d^o, d^1) as $X \twoheadrightarrow Z \rightarrowtail Y \times Y$. By the proposition and the given conditions, UZ is an equivalence relation on Y. If the diagram

$$Z \rightrightarrows Y \longrightarrow Y'$$

is a coequalizer, it is exact. Then for each $i \in I$,

$$U_i X \twoheadrightarrow U_i Z$$

and

$$U_i Z \rightrightarrows U_i Y \longrightarrow U_i Y'$$

is a coequalizer, which implies that

$$U_i X \rightrightarrows U_i Y \longrightarrow U_i Y'$$

is a coequalizer.

(6.8) Metatheorem. Let $\underline{X}$ be a regular category. Then any small diagram chasing argument valid in $\underline{S}$ is valid in $\underline{X}$, provided the data of the diagram involve only finite inverse limits and coequalizers of right exact sequences; if, moreover, the category is exact, these data may also include coequalizers of pairs of maps which, in $\underline{S}$, can be shown to have as image an equivalence relation.

(6.9) Given the somewhat vague statement of this metatheorem, it is hardly susceptible of being proved. To apply it, it is necessary only to verify that the type of diagram to be chased is by its nature susceptible of being proved by applying a family of reflexively exact functors which also reflect equivalence relations.

(6.10) **Example**. Suppose $\underline{X}$ is a regular category and we are given a commutative diagram

$$\begin{array}{ccc} Y' & \xrightarrow{f'} & X' \\ e^0 \downarrow\downarrow e^1 & & d^0 \downarrow\downarrow d^1 \\ Y & \xrightarrow[f]{} & X \\ e \downarrow & & \downarrow d \\ Y'' & \xrightarrow[f'']{} & X'' \end{array}$$

in which both columns are exact and the square

$$\begin{array}{ccc} Y' & \xrightarrow{f'} & X' \\ e^0 \downarrow & & \downarrow d^0 \\ Y & \xrightarrow[f]{} & Y \end{array}$$

is a pullback (which is equivalent to the square with e^1 and d^1 being a pullback). Then the square

$$\begin{array}{ccc} Y & \xrightarrow{f} & X \\ e \downarrow & & \downarrow d \\ Y'' & \xrightarrow[f'']{} & X'' \end{array}$$

is also a pullback.

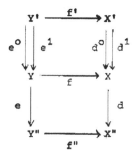

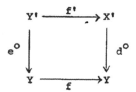

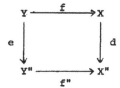

Proof. Even in the category of sets this is moderately difficult to prove. In an arbitrary regular category it follows from the metatheorem. I am indebted to Anders Kock for suggesting this example. It arises in the theory of elementary toposes and also in descent theory.

Chapter IV. Groups and Representations

1. Preliminaries.

(1.1) Throughout this chapter and the next, $\underline{X}$ denotes a fixed exact category. From I(5.11) both Gp$\underline{X}$ and Ab$\underline{X}$, the categories of groups and abelian groups in $\underline{X}$, respectively, form exact categories. The latter, in particular, is abelian.

(1.2) Let $G \in \mathrm{Gp}\underline{X}$, and $u\colon 1 \longrightarrow G$, $i\colon G \longrightarrow G$, and $m\colon G \times G \longrightarrow G$ be the unit, inverse, and multiplication maps, respectively. A pair (X,a) where $X \in \underline{X}$ and $a\colon G \times X \longrightarrow X$ is called a left representation of G or a left G-object if the following diagrams commute:

$$\begin{array}{ccc} G \times G \times X & \xrightarrow{G \times a} & G \times X \\ {\scriptstyle m \times X}\downarrow & & \downarrow{\scriptstyle a} \\ G \times X & \xrightarrow{\quad a \quad} & X \end{array} \qquad \begin{array}{ccc} X & \xrightarrow{\ \simeq\ } & 1 \times X \\ & \diagdown\!\!\diagdown & \downarrow{\scriptstyle u \times X} \\ & & G \times X \\ & & \downarrow{\scriptstyle a} \\ & & X \end{array} \quad .$$

A morphism $f\colon X \longrightarrow X'$ is a morphism of G-objects $(X,a) \longrightarrow (X',a')$ provided

$$\begin{array}{ccc} G \times X & \xrightarrow{G \times f} & G \times X' \\ {\scriptstyle a}\downarrow & & \downarrow{\scriptstyle a'} \\ X & \xrightarrow{\quad f \quad} & X' \end{array}$$

commutes.

Note that all these products exist, since, for example,

$$\begin{array}{ccc} G \times X & \longrightarrow & X \\ \downarrow & & \downarrow \\ G & \longrightarrow\!\!\!\!\!\to & 1 \end{array}$$

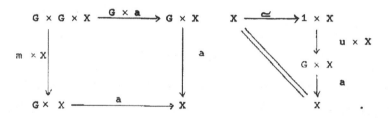

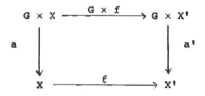

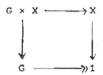

is a pullback.

The left G-objects and their morphisms evidently form a category $\underline{LO}(G)$ which has an evident underlying functor $\underline{LO}(G) \longrightarrow \underline{X}$. Turning every thing around, we can define the category $\underline{RO}(G)$ of right G-objects and their morphisms. Finally, we say that a 3-tuple (X,a,a') where $(X,a) \in \underline{LO}(G)$ and $(X,a') \in \underline{RO}(G)$ is a 2-sided G-object if

$$\begin{array}{ccc} G \times X \times G & \xrightarrow{G \times a'} & G \times X \\ {\scriptstyle a \times G}\downarrow & & \downarrow{\scriptstyle a} \\ X \times G & \xrightarrow{a'} & X \end{array}$$

commutes. The category of these objects and morphism which are simultaneously in $\underline{LO}(G)$ and $\underline{RO}(G)$ is called $\underline{BO}(G)$. It is clear that one could define G^{op} and show that $\underline{LO}(G^{op})$ is the same as $\underline{RO}(G)$ and $\underline{LO}(G \times G^{op})$ is the same as $\underline{BO}(G)$.

(1.3) Theorem. Let $\underline{X}$ be a regular category (resp. exact). Then $\underline{LO}(G)$ is regular (resp. exact) and the functor $\underline{LO}(G) \longrightarrow \underline{X}$ is a reflexively exact functor.

Proof. That it reflects isomorphisms is trivial. Now consider an exact sequence

$$X' \underset{d^1}{\overset{d^o}{\rightrightarrows}} X \xrightarrow{d} X''$$

in which (X',a') and (X,a) are left G-objects and d^o, d^1 are G-morphisms.

Then the top row of

$$\begin{array}{ccccc} G \times X' & \rightrightarrows & G \times X & \longrightarrow & G \times X'' \\ \downarrow{\scriptstyle a'} & & \downarrow{\scriptstyle a} & & \downarrow{\scriptstyle a''} \\ X' & \rightrightarrows & X & \longrightarrow & X'' \end{array}$$

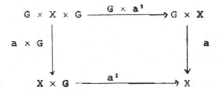

is still exact and hence a" is induced as indicated. From here the proof proceeds exactly as in I.(5.11).

(1.4) Corollary. $\underline{RO}(G)$ and $\underline{BO}(G)$ and their underlying functors to $\underline{X}$ enjoy the same properties.

Proof. This can be either proved the same way or made to follow as a corollary via the remark preceding (1.3).

(1.5) Theorem: Let $U: \underline{X} \longrightarrow \underline{Y}$ be exact. Then there is induced, for each $G \in \underline{X}$ an exact functor

$$\underline{LO}(G) \longrightarrow \underline{LO}(UG)$$

such that

$$\begin{array}{ccc} \underline{LO}(G) & \longrightarrow & \underline{LO}(UG) \\ \downarrow & & \downarrow \\ \underline{X} & \longrightarrow & \underline{Y} \end{array}$$

commutes

Proof. Recall that according to I.(5.11), UG will be a group object in $\underline{Y}$. That U takes G-objects to UG-objects follows easily from the fact that U preserves products. The exactness is a consequence of the reflexive exactness of $\underline{LO}(UG) \longrightarrow \underline{Y}$.

(1.6) Corollary. $\underline{RO}(G)$ and $\underline{BO}(G)$ enjoy the same properties.

(1.7) Lemma: Suppose (X,a,a') is an object of $\underline{BO}(G)$ and s: $G \times X \longrightarrow X \times G$ is the map which interchanges the factors. Then the immage of $G \times X \xrightarrow{(a,a'.s)} X \times X$ is an equivalence relation on X. That is, if X' is defined as the coequalizer in the diagram

$$G \times X \underset{a'.s}{\overset{a}{\rightrightarrows}} X \longrightarrow X',$$

then this sequence is right exact.

Proof. If $\underline{X}$ is small, choose $U: \underline{X} \longrightarrow \underline{S}$ which is reflexively exact and reflects equivalence relations. Then UG is an ordinary group and UX is a 2-sided UG-object. Thus it suffices to consider the case of ordinary groups operating on ordinary sets by a 2-sided operation. So we have $G \times X \longrightarrow X \times X$ by a map taking $(g,x) \longmapsto (gx,xg)$ and we want to show the image is an equivalence relation on X. It is reflexive as $(1,x) \longmapsto (x,x)$ and symmetric as $(g^{-1},gxg) \longmapsto (xg,gx)$. If (gx,xg) and $(g'x',x'g')$ satisfy $xg = g'x'$, $(gg',x'g^{-1}) \longmapsto$ $\longmapsto (gg'x'g^{-1},x'g') = (gxgg^{-1},x'g') = (gx,x'g')$, and so the image is transitive. When $\underline{X}$ is large, use an appropriate modification (cf. III.(6.4)).

2. Tensor products.

(2.1) **Proposition.** Let G be a group in $\underline{X}$, $(X,a) \in \underline{LO}(G)$ and $X' \in \underline{X}$. Then $(X \times X', a \times X') \in \underline{LO}(G)$ also.

Proof. Trivial.

(2.2) Of course $X' \times X \cong X \times X'$, so that $X' \times X \in \underline{LO}(G)$. If $(X',a') \in \underline{RO}(G)$, $X' \times X$ has the structure of a left G-object from X and of a right G-object from X'.

(2.3) **Proposition.** $X' \times X$ with this structure is an object of $\underline{BO}(G)$.

Proof. Trivial.

(2.4) **Definition.** Let $X \in \underline{LO}(G)$, $X' \in \underline{RO}(G)$. We define $X' \otimes_G X$ as the coequalizer in the diagram

$$X' \times G \times X \underset{X \times a}{\overset{a' \times X}{\rightrightarrows}} X' \times X \longrightarrow X' \otimes_G X.$$

Note that though $X' \times X$ is a left and right G-object, it is most convenient to put G in the middle. It follows from (1.7) that the sequence is right exact and thus remains right exact (in particular a coequalizer) when any right exact functor is applied.

(2.5) **Proposition.** $- \otimes_G -$ is a functor $\underline{RO}(G) \times \underline{LO}(G) \longrightarrow \underline{X}$.

Proof. If $(X,a) \xrightarrow{f} (Y,b)$ is a map of left G-objects, the diagram

$$\begin{array}{ccccc}
X' \times G \times X & \underset{X' \times a}{\overset{a' \times X}{\rightrightarrows}} & X' \times X & \longrightarrow & X' \otimes_G X \\
\Big\downarrow {\scriptstyle X' \times G \times f} & & \Big\downarrow {\scriptstyle X' \times f} & & \vdots \\
X' \times G \times Y & \underset{X' \times b}{\overset{a' \times Y}{\rightrightarrows}} & X' \times Y & \longrightarrow & X' \otimes_G Y
\end{array}$$

commutes, whence $X' \otimes f$ is induced from the coequalizer.

(2.6) Proposition. Suppose $X' \in \underline{LO}(H \times G^{op})$ (This means that it is a left H, right G, bi-object) and $X \in \underline{LO}(G)$. Then $X' \otimes_G X)$ has the natural structure of a left H object.

Proof. The top row of

$$\begin{array}{ccccc} H \times X' \times G \times X & \rightrightarrows & H \times X' \times X & \longrightarrow & H \times (X' \otimes_G X) \\ \downarrow{\scriptstyle b \times G \times X} & & \downarrow{\scriptstyle b \times X} & & \downarrow \\ X' \times G \times X & \rightrightarrows & X' \times X & \longrightarrow & X' \otimes_G X \end{array}$$

is still a coequalizer. Here $b: H \times X' \longrightarrow X'$ is, of course, the H'-structure map and the commutativity of one the squares at the left is exactly the fact of X' being a bi-object. The induced map $H \times (X' \otimes X) \longrightarrow X' \otimes X$ is easily shown to be a structure map, using, for example, that

$$H \times H \times X' \times X \longrightarrow\!\!\!\!\rightarrow H \times H \times (X' \otimes_G X).$$

(2.7) It is clear that G with its left and right multiplication maps belongs to $\underline{BO}(G)$. If $f: H \longrightarrow G$ is a morphism of group objects, there is an obvious functor $f^*: \underline{LO}(G) \longrightarrow \underline{LO}(H)$, in which $(X,a) \longmapsto (X, a.(f \times X))$. There is also included a functor $f_!: \underline{LO}(H) \longrightarrow$ $\longrightarrow \underline{LO}(G)$ which takes a H-object X to $G \otimes_H X$, evidently a G-object from the above remark.

(2.8) Theorem. The functor $f_! \dashv f^*$.

Proof. The inner adjunction is the map $X \xrightarrow{(u.t,X)} G \times X \longrightarrow G \otimes_H X$ in which $X \xrightarrow{t} I \xrightarrow{u} G$ is the terminal map of X followed by the unit of G. The outer adjunction is induced by

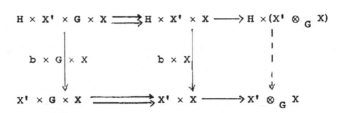

$$\begin{array}{ccccc} G \times H \times X & \rightrightarrows & G \times X & \longrightarrow & G \otimes_H X \\ & & \Big\downarrow a & \swarrow & \\ & & X & & \end{array}$$

That the first is H linear, the second exists and is G-linear, and the two satisfy the laws of an adjunction may be easily verified by applying the metatheorem.

(2.9) Corollary. For any G, the underlying functor $\underline{BO}(G) \longrightarrow \underline{X}$ has a left adjoint, $X \longmapsto G \times X$.

Proof. Apply the above to $G \longrightarrow 1$. It is evident that $G \otimes_1 X = G \times X$.

(2.10) Theorem. Let $X \in \underline{LO}(G \times H^{op})$, $Y \in \underline{LO}(H \times K^{op})$, $Z \in \underline{LO}(K \otimes L^{op})$. Then there is a canonical map

$$(X \otimes_H Y) \otimes_K Z \longrightarrow X \otimes_H (Y \otimes_K Z)$$

such that the diagram

$$\begin{array}{ccc} & X \times Y \times Z & \\ \swarrow & & \searrow \\ (X \otimes_H Y) \otimes_K Z & \longrightarrow & X \otimes_H (Y \otimes_K Z) \end{array}$$

commutes (see the proof for the definition of these vertical maps), and that map is an isomorphism.

Proof. The vertical maps in the diagram are gotten by letting $t(X,Y)$ denote the canonical projection $X \times Y \longrightarrow X \otimes_H Y$. Then the one map is $t(X \otimes_H Y, Z) . t(X,Y) \otimes Z$ and the other is similar. One way of proving this is to first prove it in $\underline{S}$ (trivial). Then use the metatheorem to show that in the diagram

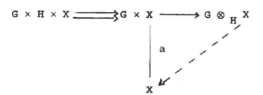

$$\begin{array}{ccccc} X \times Y \times K \times Z & \rightrightarrows & X \times Y \times Z & \longrightarrow & X \times (Y \otimes_K Z) \\ & & \downarrow & \swarrow & \\ & & (X \otimes_H Y) \otimes_K Z & & \end{array}$$

the vertical arrow coequalizes the two maps on the left. Since the row is a right exact, it is a coequalizer, and there is induced $X \times (Y \otimes_K Z) \longrightarrow (X \otimes_H Y) \otimes_K Z$ with the appropriate property. Another use of the metatheorem shows that in the diagram

$$\begin{array}{ccccc} X \times H \times (Y \otimes_K Z) & \rightrightarrows & X \times (Y \otimes_K Z) & \longrightarrow & X \otimes_H (Y \otimes_K Z) \\ & & \downarrow & \swarrow & \\ & & (X \otimes_H Y) \otimes_K Z & & \end{array}$$

the vertical arrow again coequalizes the two arrows on the left and the required map is the one induced. That it is an isomorphism may be readily verified by a third use of the embedding.

(2.11) <u>Theorem</u>: If $X \in \underline{LO}(G)$, $G \otimes_G X \cong G$; and if $Y \in \underline{RO}(G)$, $Y \otimes_G G \cong Y$.

Proof. These can be derived either directly from adjointness or from arguments similar to (but simpler than) the above.

(2.12) <u>Theorem</u>: The associativity and unit of the previous two theorems are jointly coherent.

Proof. Prove it in $\underline{S}$ and use the metatheorem.

(2.13) <u>Corollary</u>. If $g: K \longrightarrow H$, $f: H \longrightarrow G$, then $(f.g)_! = f_! . g_!$.

Proof. From the previous theorems we have for $X \in \underline{LO}(K)$, $f_!(g_!X) = G \otimes_H (H \otimes_K X) \cong (G \otimes_H H) \otimes_K X \cong G \otimes_K X = (fg)_!(X)$.

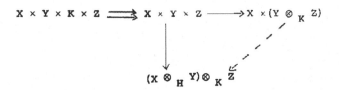

(2.14) Remark. Later on, when G is commutative (and then $\underline{LO}(G)$ and $\underline{RO}(G)$ are equivalent to the same full subcategory of $\underline{BO}(G)$, namely the subcategory of symmetric objects), there will be a commutativity isomorphism as well, which by the same reasoning will be jointly coherent with the above.

(2.15) Proposition. Let $U: \underline{X} \longrightarrow \underline{Y}$ be an exact functor, $G \in \underline{X}$, $X_1 \in \underline{RO}(G)$, and $X_2 \in \underline{LO}(G)$. Then

$$U(X_1 \otimes_G X_2) \cong UX_1 \otimes_{UG} UX_2.$$

Proof. Exact functors preserve both products and right exact sequences. Apply U to

$$X_1 \times G \times X_2 \rightrightarrows X_1 \times X_2 \longrightarrow X_1 \otimes_G X_2.$$

3. Principal objects.

(3.1) Definition. Let G be a group in $\underline{X}$. A left G-object X will be called a principal left G-object if

a) $X \longrightarrow\!\!\!\!\rightarrow 1$.

b) $G \times X \xrightarrow{(a,p_2)} X \times X$ is an isomorphism. Here $a: G \times X \longrightarrow X$ is the structure while $p_2: G \times X \longrightarrow X$ is the second coordinate projection. We let $\underline{PLO}(G)$ denote the full subcategory of these objects.

(3.2) The definition is, in view of III(2.11), exactly the same as Chase's [Ch] which goes back, in turn, to Beck [Be]. Much of the preliminary material in this section is special cases of results proved by Chase, His proofs, however, were generally much more complicated because he had no metatheorem available.

(3.3) Proposition. Let $U: \underline{X} \longrightarrow \underline{Y}$ be exact. Then $U(\underline{PLO}(G)) \subset \underline{PLO}(UG)$.

Proof. U preserves $\longrightarrow\!\!\!\!\rightarrow$, finite products, and (like any functor) isomorphisms.

(3.4) Proposition. Let G be a group (in $\underline{S}$). Then $\underline{PLO}(G)$ consists (up to isomorphism) of the single object G, and the morphisms, all $\xrightarrow{\sim}$, consist of the right multiplications by the elements of G.

Proof. Let $X \in \underline{PLO}(G)$. Condition i) of (3.1) says that $X \neq \emptyset$. Condition ii) says that the map $G \times X \longrightarrow X \times X$, which takes $(g,x) \longmapsto (gx,x)$ for $g \in G$ and $x \in X$, is an isomorphism. This amounts to saying that if x is held fixed, there is for each $x' \in X$ a unique solution in G to $gx = x'$. In other words, if $x \in X$ is fixed, the mapping $G \longrightarrow X$ by $g \longmapsto gx$ is an isomorphism. The rest of the proposition is trivial.

(3.5) Proposition. $\underline{PLO}(G)$ is a groupoid (that is every map is $\xrightarrow{\sim}$).

Proof. If $X \longrightarrow X'$ is a map in $\underline{PLO}(G)$ choose an embedding and apply the last proposition.

(3.6) Proposition $X \in \underline{PLO}(G)$ is isomorphic to G if and only if there is a map $1 \longrightarrow X$ in $\underline{X}$. In fact, $\underline{PLO}(G)(G,X) \cong \underline{X}(1,X)$.

Proof. $\underline{PLO}(G) \subset \underline{LO}(G)$ is full and faithful. Hence this follows from adjointness:

$$\underline{LO}(G)(G,X) = \underline{LO}(G)(G \times 1,X) \cong \underline{X}(1,X).$$

(3.7) Theorem: Let $U: \underline{X} \longrightarrow \underline{S}$ range over a family of exact embeddings which collectively reflect isomorphisms. Then $\underline{PLO}(G)$ consists of those X for which $UX \cong UG$ as UG-objects.

Proof. If $UX = UG$, then the canonical map $(Ua,p_2): UG \times UX \longrightarrow UX \times UX$ is an isomorphism, which means that $U(a,p_2): U(G \times X) \longrightarrow U(X \times X)$ is also, and finally that $(a,p_2): G \times X \longrightarrow X \times X$ is. On the other hand, by (3.3) and (3.4), $X \in \underline{PLO}(G)$ implies $UX \cong UG$.

(3.8) Theorem: Let $f: H \longrightarrow G$ be a morphism of groups. Then $f_!(\underline{PLO}(H)) \subset \underline{PLO}(G)$.

Proof. For any exact $U: \underline{X} \longrightarrow \underline{S}$, $U(G \otimes_H X) \cong UG \otimes_{UH} UX \cong UG \otimes_{UH} UH \cong UG$. Note that $f_!$ is not in general exact, so that (3.3) does not apply here.

(3.9) Proposition. Suppose $f: H \longrightarrow G$ is the trivial map, $H \longrightarrow 1 \xrightarrow{u} G$. Then for $X \in \underline{PLO}(H)$, $f_!(X) \cong G$.

Proof. It is sufficient to show that there is a G-morphism of $f_!(X) \longrightarrow G$. In the diagram

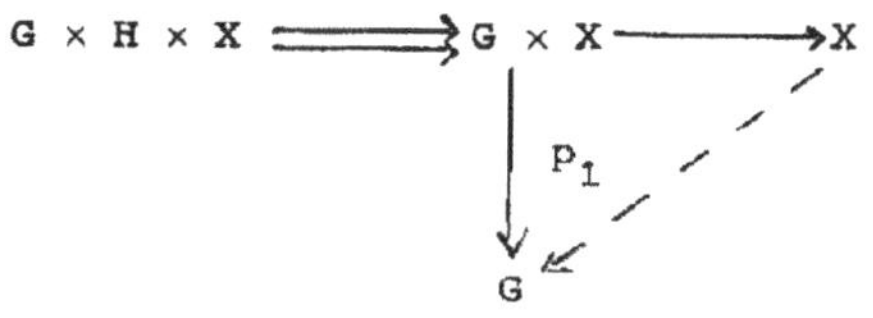

the vertical map coequalizes the two maps on the left (the structure $G \times H \longrightarrow G$, is in this case just the projection) and induces $X \longrightarrow G$, evidently a G-morphism.

4. Structure of groups.

(4.1) In this section we derive a few results about the relation between kernels and kernel pairs. We continue to let $\underline{X}$ denote an exact category.

(4.2) We know from I.(5.11) that the underlying functor from Gp $\underline{X} \longrightarrow \underline{X}$ is exact and hence preserves limits and regular epimorphisms. Since the category is also pointed, the notions of normal monomorphisms and epimorphisms also arise. It is evident that a normal epimorphism is always regular, but in general (e.g. in pointed sets) the converse is not always true. Here we will show that it is.

(4.3) Proposition. Gp $\underline{X}$ has finite products.

Proof. The terminal map $G \longrightarrow 1$ of any group is $\longrightarrow\!\!\!\!\!\rightarrow$, being split by the unit. Then the pullback

$$\begin{array}{ccc} G \times H & \twoheadrightarrow & G \\ \downarrow & & \downarrow \\ H & \twoheadrightarrow & 1 \end{array}$$

exists.

(4.4) Proposition. Gp $\underline{X}$ has finite limits.

Proof. It is necessary only to show that equalizers exist. During this argument we will denote the composition of morphisms by a dot, as f.g, while the multiplication of two morphisms to some group will be denoted simply by juxtaposition, as fg. The inverse, under the group law, will be denoted f^{-1}. This latter is particularly ambiguous but none of the maps arising in the proof will be isomorphisms (except accidently) and the inverse in the category will not be used. Of course neither f^{-1} nor fg will generally be morphisms of Gp $\underline{X}$ when f and g are. Now suppose we are given two maps $f,g\colon G \longrightarrow H$. We let

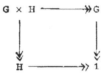

$u\colon 1 \longrightarrow G$, $1 \longrightarrow H$ denote interchangeably the unit morphisms. In particular $f.u = u$, $g.u = u$ and $fg^{-1}.u = (f.u)(g^{-1}.u) = (f.u)(g.u)^{-1}) = = uu^{-1} = uu = u$. If X is the image of $fg^{-1}\colon G \longrightarrow H$, this shows that $u\colon 1 \longrightarrow H$ factors through X via fg^{-1}. Now let K be the pullback in the diagram

$$\begin{array}{ccccc} K & \xrightarrow{t} & 1 & & \\ {\scriptstyle h}\downarrow & & \downarrow & \searrow^{u} & \\ G & \twoheadrightarrow & X & \longrightarrow & H \end{array}$$

Once this pullback exists, it follows that

$$\begin{array}{ccc} K & \longrightarrow & 1 \\ {\scriptstyle h}\downarrow & & \downarrow \\ G & \xrightarrow{fg^{-1}} & H \end{array}$$

is also a pullback.

Now K is a group, and in particular $h\colon K \rightarrowtail G$ is a subgroup, if and only if $(X,K) \xrightarrow{(X,h)} (X,G)$ is a subgroup for each X. Applying $(X,-)$, we still get a pullback in $\underline{S}$

$$\begin{array}{ccc} (X,K) & \longrightarrow & (X,1) = 1 \\ \downarrow & & \downarrow \\ (X,G) & \xrightarrow[(X,f)(X,G)^{-1}]{} & (X,H) \end{array}$$

and (X,K) really is the equalizer of the two group homomorphisms (X,f) and (X,g), and hence is a subgroup.

(4.5) Proposition. Every regular epimorphism is normal.

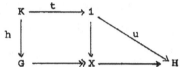

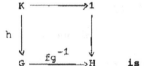

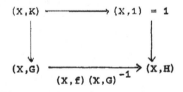

Proof. We use the same conventions as in the proof above. The underlying functor $\text{Gp}\,\underline{X} \longrightarrow \underline{X}$ preserves finite inverse limits. It preserves, in particular, kernels, since the kernel of a map is the equalizer of that map and the trivial map. As in (3.9), we let u also denote this trivial map between any two groups. Now suppose that

$$G' \underset{e}{\overset{d}{\rightrightarrows}} G \xrightarrow{f} G''$$

is a coequalizer and $H \xrightarrow{g} G$ is the kernel of f. We want to show that f is the cokernel of h, and it clearly suffices to show that for any h: $G \longrightarrow K$, h.g = u implies h.e = h.d. But g is also the equalizer of f and u as maps in $\underline{X}$. Now $f.de^{-1} = (f.d)(f.e^{-1}) = (f.d)(f.e)^{-1} =$ $=(f.d)(f.d)^{-1} = u$. Hence there is map k: $G' \longrightarrow H$ such that $g.k = de^{-1}$. Now for any h: $G \longrightarrow K$ with $h.g = u$, $u = h.g.k = h.de^{-1} =$ (as above) $(h.d)(h.e)^{-1}$, and on multiplying this by u, which is the unit of (G,K), we have h.e = h.d, which completes the proof.

Chapter V. Cohomology.

1. Definitions.

(1.1) In this chapter we will define cohomology sets of $\underline{X}$ with coefficients in a group in $\underline{X}$. Only H^0 and H^1 will be defined here. There are several suggestions for higher sets; these are being investigated currently. The "cohomology sets" are covariant functors of the coefficients. What they are contravariant functors of is suggested by the classical examples (cf. section 4). If $\underline{X}$ is exact, so is $(\underline{X},X)$ for any $X \in \underline{X}$ by I.(5.4); and if $X \longrightarrow X'$ is a map, there is induced $(\underline{X},X') \longrightarrow (\underline{X},X)$ by pulling back, provided the pullbacks exist. Even if they don't, they do for all $Y \twoheadrightarrow X'$, and that is all the cohomology is concerned with. If G is a group in $(\underline{X},X)$, it also is in $(\underline{X},X')$, and there is induced $H^i(X',G) \longrightarrow H^i(X,G)$, $i = 0,1$. In the discussion below, the X is suppressed and we write $H^i(G)$, which should actually be $H^i(1,G)$. (X is terminal in $(\underline{X},X)$ and the cohomology of X is the cohomology of that terminal object.)

(1.2) Throughout this chapter we will keep certain notational conventions. In addition to $\underline{X}$ being exact, we suppose that it has a terminal object 1 and that $t: X \longrightarrow 1$ denotes the terminal map of every object. Each group comes equipped with its multiplication m, its inverse i, and its unit u. For any object X and group G, we will also use $u: X \longrightarrow G$ to denote the composite $X \xrightarrow{t} 1 \xrightarrow{u} G$. The maps denoted t form a right ideal with respect to all the objects and those denoted by u form a left ideal with respect to groups and group homomorphisms. In addition, for this section we fix an exact sequence of groups and group homomorphisms

$$1 \xrightarrow{u} G' \xrightarrow{f} G \xrightarrow{f'} G'' \xrightarrow{t} 1.$$

(1.3) The cohomology will be relative to an underlying functor U: $\underline{X} \longrightarrow \underline{Y}$. Although the functor U and the category $\underline{Y}$ are usually exact, it seems desirable to develope the relative theory without those assumptions. Accordingly we will suppose only that U preserves finite limits. The absolute, or unrelativized, theory may be recovered by letting U be an exact functor to a category $(\underline{C},\underline{S})$ where $\underline{C}$ is discrete, for in that category every epimorphism splits and every principal G-object is isomorphic to G. The desirability of considering such a relative theory was pointed out by Jon Beck.

(1.4) Definition. Let G be a group in $\underline{X}$ and $X \in \underline{PLO}(G)$. We say that X is split by a functor U if $UX \cong UG$ as a UG object.

(1.5) Proposition. With U,X and G as above, X is split by U if and only if there is a morphism $1 \longrightarrow UX$.

Proof. Of course in the case in which $\underline{Y}$ is exact, this follows from IV.(3.6). But we have not supposed that. In any event, $(1,UG) \neq \emptyset$, so one direction is trivial. To go the other way, let $H = UG$ and $Y = UX$, and suppose there is a map $s: 1 \longrightarrow Y$. Now H is a group, Y is an H object, and $H \times Y \xrightarrow{\sim} Y \times Y$. This implies that the representable functor $(-,H)$ is a group, $(-,Y)$ is an H-object, and

$$(-,H) \times (-,Y) \xrightarrow{\sim} (-,Y) \times (-,Y).$$

Then for any Y' such that $(Y',Y) \neq \emptyset$, (Y',Y) is a principal (Y',G) object. This implies that $(Y',G) \xrightarrow{\sim} (Y',Y)$ by the map that, associates to a fixed $f_o: Y' \longrightarrow Y$ and to an arbitrary map $g: Y' \longrightarrow G$, the map

$$Y' \xrightarrow{(g,f_o)} G \times Y \longrightarrow Y,$$

the second map being the structure. If we take for f_o the composite

$$Y' \xrightarrow{t} 1 \xrightarrow{s} Y,$$

this defines a natural $(-,G)$ equivalence $(-,G) \xrightarrow{\sim} (-,Y)$ which must be induced by a G equivalence $G \xrightarrow{\sim} Y$.

(1.6) <u>Definition.</u> We know that <u>PLO</u>(G) is a groupoid (IV.(3.5)). In addition, there is a distinguished component in <u>PLO</u>(G), the one containing G. We define H^0G to be the set of automorphisms of G, and given $U: \underline{X} \longrightarrow \underline{Y}$, we define $H^1(U,G)$ to be the set - or maybe class - of all components of <u>PLO</u>(G) split by U. That means those components containing a representative split by U. Since the distinguished component is clearly split by U, this may be considered as a pointed set - or class - with the distinguished component as base point. In the case that the functor U is exact and takes values in <u>S</u>, whence every $X \in$ <u>PLO</u>(G) splits, the resultant set $H^1(U,G)$ is simply the set of connected components of <u>PLO</u>(G) and is denoted $H^1(G)$. This is the "absolute" cohomology.

(1.7) <u>Proposition</u>. Let $f: G' \longrightarrow G$ be a group homomorphism. Then if $X \in$ <u>PLO</u>(G') is U split, so is $f_!(X) \in$ <u>PLO</u>(G).

Proof. There is a map $X \longrightarrow f_!(X)$ (essentially the front adjunction) and a map $1 \longrightarrow UX$ gives one $1 \longrightarrow UX \longrightarrow Uf_!(X)$.

(1.8) <u>Theorem</u> (Beck). Suppose <u>X</u> is exact and $U: \underline{X} \longrightarrow \underline{Y}$ is a tripleable underlying functor. Then for $G \in \mathrm{Gp}\,\underline{X}$, $H^0(G)$ and $H^1(U,G)$ are the zeroth and first (non-abelian) triple cohomology sets of the object 1 with coefficients in G.

The proof is rather long and is given in [Be]. If F is left adjoint to U and the front and back adjunctions are given by $\eta: \underline{Y} \longrightarrow UF$ and $\varepsilon: FU \longrightarrow \underline{X}$, then the triple sets are computed from the complex

$$1 \longrightarrow \underline{X}(FU1,G) \rightrightarrows \underline{X}(FUFU1,G) \overset{\longrightarrow}{\rightrightarrows} \underline{X}(FUFUFU1,G),$$

the arrows induced by such things as εFU and $FU\varepsilon$ and similar maps at the next stage. The fact, standard in tripleable categories, that

$$FUFUX \underset{UF\varepsilon X}{\overset{FU\varepsilon X}{\rightrightarrows}} FUX \xrightarrow{\varepsilon X} X$$

is a coequalizer, implies easily, if X is taken as 1, that the zeroth cohomology is $\underline{X}(1,G)$.

(1.9) Corollary. Suppose $U\colon \underline{X} \longrightarrow \underline{S}$ is tripleable. Then U is exact and the zeroth and first triple cohomology of the object 1 with coefficients in a group object G are exactly $H^{0}(G)$ and $H^{1}(G)$.

Proof. The exactness of U in this case is well-known (in fact is the direct ancestor of the definition of exactness used in this paper) and the rest then follows from the preceding theorem.

2. The exact sequence.

(2.1) If $U\colon \underline{X} \longrightarrow \underline{Y}$ is a finite limit preserving functor and

$$1 \longrightarrow G' \xrightarrow{f} G \xrightarrow{f'} G'' \longrightarrow 1$$

is an exact sequence in Gp $\underline{X}$, we say that it is a U-split exact sequence if Uf' is a split epimorphism. Thus

$$1 \longrightarrow UG' \xrightarrow{Uf} UG \xrightarrow{Uf'} UG'' \longrightarrow 1$$

is a split exact sequence.

(2.2) Theorem. Let $U\colon \underline{X} \longrightarrow Y$ preserve finite limits and

$$1 \longrightarrow G' \longrightarrow G \longrightarrow G'' \longrightarrow 1$$

be a U-split exact sequence. Then there is a natural map $\delta\colon H^{o}G'' \longrightarrow H^{1}(U,G')$ such that the resulting sequence

$$\begin{array}{l} 1 \longrightarrow H^{o}G' \longrightarrow H^{o}G \longrightarrow H^{o}G'' \\ \swarrow \\ H^{1}(U,G') \longrightarrow H^{1}(U,G) \longrightarrow H'(U,G'') \end{array}$$

is exact, the last four terms being exact as a sequence of pointed sets.

Proof. One can easily show that $1 \longrightarrow G' \longrightarrow G \longrightarrow G''$ being an exact sequence in Gp $\underline{X}$ is equivalent to

$$1 \longrightarrow (-,G') \longrightarrow (-,G) \longrightarrow (-,G'')$$

being an exact sequence of group valued functors on $\underline{X}$ (cf. I.(5.10)). In particular, evaluated at 1, we get

$$1 \longrightarrow (1,G') \longrightarrow (1,G) \longrightarrow (1,G'')$$

is exact, which gives the exactness of half of the sequence. The next step is to give the connecting map. Suppose $d\colon 1 \longrightarrow G''$ is given (we identify $(1,G'')$ with Aut G''). Let X be the pullback in the diagram

$$\begin{array}{ccc} X & \xrightarrow{t} & 1 \\ {\scriptstyle g}\downarrow & & \downarrow{\scriptstyle d} \\ G & \longrightarrow & G'' \end{array} .$$

Since $G \longrightarrow G''$ is a U-split epimorphism and U preserves pullback, $X \longrightarrow 1$ is also a U-split epimorphism. A map

$$a: G' \times X \longrightarrow X$$

is defined by $t.a = t$ and $g.a = (f.p_1)(g.p_2)$. Recall that t denotes everybody's terminal map, p_1 and p_2 are coordinate projections, and q.a is to be the product in the group $\underline{X}(G' \times X,G)$ of $(f.p_1)$ and $(q.p_2)$ We see that a is well defined from

$f'.(f.p_1)(q.p_2) = (f'.f.p_1)(f'.q.p_2) = (u.p_1)(d.t.p_2) = u(d.t) = d.t.$

Here we use the fact that f' is a homomorphism of group objects. To see that this gives X the structure of a a principal G-object — evidently U-split — it suffices to consider the situation in $\underline{S}$. There d picks out a point of G" and X is the inverse image of that point, operated on by left translation by G'. It is evidently isomorphic to G' in that case and so, in general, is a principal G'-object whose class we denote by $\delta(d)$.

(2.3) Proposition. The sequence

$$H^0G \longrightarrow H^0G'' \longrightarrow H^1(U,G')$$

is exact.

Proof. Refering to the definition of $\delta(d)$ above, we see that if d lifts to a map $1 \longrightarrow G$, this gives a splitting of $X \longrightarrow 1$ by the pullback property. The converse is trivial.

(2.4) Proposition. The sequence

$$H^0G'' \longrightarrow H^1(U,G') \longrightarrow H^1(U,G)$$

is exact.

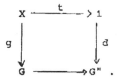

Proof. If $d: 1 \longrightarrow G''$ is given, and X is a principal G'-object representing $\delta(d)$, X comes equipped with a map $X \xrightarrow{q} G$, easily seen to be G'-linear. From the adjointness

$$\mathrm{Hom}_{G'}(X,G) \xrightarrow{\sim} \mathrm{Hom}_G(G \otimes_{G'} X, G)$$

we see that there is a map $G \otimes_{G'} X \longrightarrow G$ and so they are isomorphic. Conversely, if they are isomorphic, there is a map $X \xrightarrow{q} G$. Consider the diagram

$$\begin{array}{ccccc} G' \times X & \overset{a}{\underset{p_2}{\rightrightarrows}} & X & \longrightarrow & 1 \\ & & \Big\downarrow q & & \\ & & G & \xrightarrow{f'} & G''. \end{array}$$

Since $(a,p_2): G' \times X \xrightarrow{\sim} X \times X$ and $X \twoheadrightarrow 1$, the top row is a coequalizer. The facts that $f'.f = u$ and q is a G'-linear morphism imply that $f'.q.a = f'.q.p_2$ (e.g., use the metatheorem) and hence a map $d: 1 \longrightarrow G''$ is induced making the square commute. If $\delta(d)$ is represented by an $X' \in \underline{PO}(G')$, the properties of pullback give a map $X \longrightarrow X'$, easily seen to be a G-morphism and hence an isomorphism.

(2.5) Proposition. The sequence

$$H^1(U,G') \longrightarrow H^1(U,G) \longrightarrow H^1(U,G'')$$

is exact.

Proof. The composite map is $f_!.f'_! = (f'.f)_! = u_!$, which is trivial by IV(3.9). To go the other way, suppose that $G'' \otimes_G X \cong G''$. The front adjunction gives a map $X \longrightarrow G'' \otimes_G X$ and we see from the commutative diagram

$$\begin{array}{ccc} G \times X & \longrightarrow & G' \times X \\ \downarrow & & \downarrow \\ X & \longrightarrow & G' \otimes_G X \end{array}$$

that $X \twoheadrightarrow G' \otimes_G X$ Then we may pull this back along any $1 \longrightarrow G' \otimes_G X$ to obtain

$$\begin{array}{ccc} X' & \twoheadrightarrow & 1 \\ \downarrow & & \downarrow \\ X & \twoheadrightarrow & G' \otimes_G X \end{array} .$$

The map $G' \times X' \longrightarrow G \times X \longrightarrow X$ gives X' the structure of a G' object. Applying U, we get a pullback square

$$\begin{array}{ccc} UX' & \longrightarrow & 1 \\ \downarrow & & \downarrow \\ UG & \longrightarrow & UG'' \end{array} .$$

Since $UG \longrightarrow UG''$ is a split epimorphism, so is $UX' \longrightarrow 1$. Similarly, we may use the metatheorem to see that $X' \in$ PLO(G'). Finally, the map $X' \longrightarrow X$, easily seen to be a G'-morphism, gives a G-isomorphism $G \otimes_{G'} X' \xrightarrow{\sim} X$. This completes the proof of (1.2).

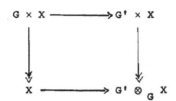

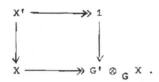

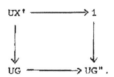

3. Abelian groups.

(3.1) In this section we consider the special case of the theorem (2.2) in which G is abelian. To emphasize this fact, we use A instead of G throughout this section to denote an abelian group object of X. Ab X denotes the category of abelian group objects of X and morphisms of groups. The first observation we have is an immediate consequence of I.(3.11) and I.(5.11).

(3.2) Theorem: Let X be an exact category. Then Ab X is abelian.

(3.3) When A is abelian LO(A) can be embedded as a full subcategory of BO(A) as the subcategory of symmetric objects. Namely, given an a: $A \times X \longrightarrow X$ making X into a left A-object, X becomes a right A-object, indeed a 2-sided A-object, via the composite

$$X \times A \longrightarrow A \times X \xrightarrow{a} X,$$

in which the first morphism is the switching isomorphism. Via this embedding we may consider the tensor product as defining a functor

$$- \otimes - : \underline{LO}(A) \times \underline{LO}(A) \longrightarrow \underline{BO}(A).$$

(3.4) Proposition. The image of the isomorphism above is contained in LO(A).

Proof. In sets, a symmetric 2-sided A-object X satisfies $ax = xa$. In $X \otimes_A Y$, we have $a(x \otimes y) = ax \otimes y = xa \otimes y = x \otimes ay = x \otimes ya = (x \otimes y)a$, given that both X and Y are symmetric. Now use the meta-theorem.

(3.5) Proposition. The image of $- \otimes -$ restricted to $\underline{PLO}(A) \times \underline{PLO}(A)$ is contained in PLO(A).

Proof. Using IV.(2.11), IV.(2.15) and IV.(3.7), we have, for X, $Y \in \underline{PLO}(A)$, and for exact U: $\underline{X} \longrightarrow \underline{S}$,

$$U(X \otimes_A Y) \cong UX \otimes_{UA} UY \cong UA \otimes_{UA} UA \cong UA,$$

whence by again applying IV.(3.7) $X \otimes_A Y \in \underline{PLO}(A)$.

(3.6) *Proposition*. The functor $- \otimes_A - : \underline{LO}(A) \times \underline{LO}(A) \longrightarrow \underline{LO}(A)$ is associative, commutative, and unitary up to jointly coherent isomorphism.

Proof. Prove it in $\underline{S}$ and use the metatheorem.

(3.7) *Corollary*. The set $H^1(A)$ is an abelian monoid, the product being induced by $- \otimes_A -$.

(3.8) *Theorem*. $H^1(A)$ is an abelian group with respect to the tensor product.

Proof. We need only show that there are inverses. Let $X \in \underline{LO}(G)$ have structure map $a: A \times X \longrightarrow X$ and $i: A \longrightarrow A$ be the inverse map of A, a homomorphism since A is commutative. Let $X^{\#}$ denote X with structure map

$$A \times X \xrightarrow{\; i \times X \;} A \times X \xrightarrow{\; a \;} X.$$

An application of the embedding shows that it is principal. Let b: $X \times X \longrightarrow A$ be the composite

$$X \times X \xrightarrow{\; (a,p_2)^{-1} \;} A \times X \xrightarrow{\; p_1 \;} A$$

from which $(a,p_2)^{-1} = (b,p_2)$. Now consider

$$\begin{array}{ccccc} X \times A \times X^{\#} & \rightrightarrows & X \times X^{\#} & \longrightarrow & X \otimes_A X^{\#} \\ & & \Big\downarrow b & \swarrow_{\dashrightarrow} & \\ & & A & & \end{array}$$

which makes sense since X and $X^{\#}$ are the same object of $\underline{X}$. In sets,

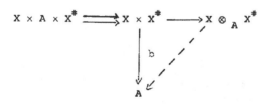

$A \cong X$, and we may suppose $A = X$. In that case, $a\colon A \times A \longrightarrow A$ is addition and we may easily check that $b\colon A \times A \longrightarrow A$ is subtraction, $p_1 - p_2$. Then b coequalizes the two maps $X \times A \times X^{\#}$ to $X \times X^{\#}$. Then there is induced a map $X \otimes_A X^{\#} \longrightarrow A$, easily seen to be an A-morphism, hence an isomorphism. The metatheorem allows us to pull this argument back to $\underline{X}$.

(3.9) __Proposition__. If $U\colon \underline{X} \longrightarrow \underline{Y}$ preserves finite limits, $H^1(U,A)$ is a subgroup of $H^1(A)$.

Proof. If UX_1 and UX_2 are split, then we have a map $1 \longrightarrow UX_1 \times UX_2 \cong U(X_1 \times X_2) \longrightarrow U(X_1 \otimes_A X_2)$, the latter being this image under U of the natural projection $X_1 \times X_2 \longrightarrow X_1 \otimes_A X_2$. If $X \otimes_A X^{\#} \cong A$, then X and $X^{\#}$ are isomorphic in $\underline{X}$, so UX splits if and only if $U(X^{\#})$ does. Finally, the trivial class, that of A, splits already in $\underline{X}$.

(3.10) __Theorem__: Let $U\colon \underline{X} \longrightarrow \underline{Y}$ preserve finite limits and $0 \longrightarrow A' \longrightarrow A \longrightarrow A'' \longrightarrow 0$ be a U-split exact sequence in Ab $\underline{X}$. Then the sequence of (2.2) is an exact sequence of abelian groups.

Proof. $0 \longrightarrow H^0(A') \longrightarrow H^0(A) \longrightarrow H^0(A'')$ is obviously exact in $\underline{Ab}$. For $g\colon B \longrightarrow B'$, the induced map $H^1(U,B) \longrightarrow H^1(U,B_1)$ is given by $X \longmapsto B' \otimes_B X$. Using (3.6), we have $(B' \otimes_B X_1) \otimes_{B'} (B' \otimes_B X_2)$ $\cong ((B' \otimes_B X_1) \otimes_{B'} B') \otimes_B X_2) \cong (B' \otimes_B X_1) \otimes_B X_2 = B' \otimes_B (X_1 \otimes_B X_2)$ so that the induced map $H^1(U,B) \longrightarrow H^1(U,B')$ is an abelian group homomorphism. In particular

$$H^1(U,A') \longrightarrow H^1(U,A) \longrightarrow H^1(U,A'')$$

is an exact sequence of abelian groups. Thus we need only show that the connecting homomorphism $\delta\colon H^0(A'') \longrightarrow H^1(U,A')$ is additive. That is, given

$$\begin{array}{ccc} X_1 & \longrightarrow & 1 \\ {\scriptstyle q_1}\downarrow & & \downarrow{\scriptstyle d_1} \\ A & \longrightarrow & A'' \end{array} \qquad\qquad \begin{array}{ccc} X_2 & \longrightarrow & 1 \\ {\scriptstyle q_2}\downarrow & & \downarrow \\ A & \longrightarrow & A'' \end{array}$$

pullback squares, we must show that there is a pullback square

$$\begin{array}{ccc} X_1 \otimes_A X_2 & \longrightarrow & 1 \\ \downarrow & & \downarrow{\scriptstyle d_1 + d_2} \\ A & \longrightarrow & A'' \end{array}$$

As in the proof of (1.10), it is sufficient merely to exhibit a commutative square of that sort. Consider the diagram

$$\begin{array}{ccccc} X_1 \times A' \times X_2 & \rightrightarrows & X_1 \times X_2 & \longrightarrow & X_1 \otimes_{A'} X_2 \\ & & \downarrow{\scriptstyle q_1 \times q_2} & & \\ & & A \times A & & \\ & & \downarrow{\scriptstyle m} & & \\ & & A & & \end{array}$$

where m is the addition. By applying the metatheorem we see that the vertical map coequalizes the given maps and induces $X_1 \otimes_{A'} X_2 \longrightarrow A$. Another application of the embedding (or a simple direct argument based on the facts that m induces the addition in $(-,A)$ and that $A \longrightarrow A''$ is a homomorphism) shows that

$$\begin{array}{ccc} X_1 \otimes X_2 & \longrightarrow & 1 \\ \downarrow & & \downarrow{\scriptstyle d_1 + d_2} \\ A & \longrightarrow & A'' \end{array}$$

commutes.

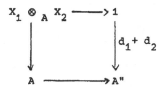

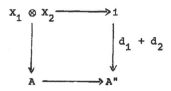

4. Extensions.

(4.1) Consider an exact category $\underline{X}$ and a fixed object X. Then $\underline{Y} = (\underline{X}, X)$ is also exact by I.(5.4). This category also has a terminal object, $X \longrightarrow X$, by the identity map. A map $Y \longrightarrow\!\!\!\!\rightarrow X$ will be called an extension of X. If G is a group of $\underline{Y}$, we say that G is an X-group. A principal G-object is a $Y \longrightarrow\!\!\!\!\rightarrow X$ on which G operates principally. It is in particular an extension and will be called a singular extension with kernel G. $G \longrightarrow\!\!\!\!\rightarrow X$ itself will be called the split extension with kernel G. Note that the unit law shows up in this case as a map $X \longrightarrow G$ which splits $G \longrightarrow\!\!\!\!\rightarrow X$, so that this really is a split epimorphism. In particular, a U-split extension is one which really splits when U is applied.

(4.2) Suppose $\underline{X}$ is the category $\underline{Gp}$ of groups and $X \in \underline{X}$ is a fixed group. Then an X-group G is a $G \longrightarrow\!\!\!\!\rightarrow X$ whose group law considered as a map $G \times_X G \longrightarrow G$ is a homomorphism of groups. Since $G \longrightarrow\!\!\!\!\rightarrow X$ is split, G is a semi-direct product $X \times M$ where M is the kernel of $G \longrightarrow X$. $G \times_X G$ is $X \times M \times M$ and it is a moment's calculation to see that M must be abelian and that G operates on M as a G-module.

(4.3) If

$$0 \longrightarrow M \longrightarrow G \longrightarrow X \longrightarrow 1$$

and

$$0 \longrightarrow M \longrightarrow Y \longrightarrow X \longrightarrow 1$$

are (still in the category of groups) two singular extensions of X with kernel M, the upper being split, then we can form the pullbacks

$$\begin{array}{llcccccccl}
(1) & 0 & \longrightarrow & M & \longrightarrow & G & \longrightarrow & X & \longrightarrow & 1 \\
 & & & \| & & \uparrow{\scriptstyle p_1} & & \uparrow & & \\
(2) & 0 & \longrightarrow & M & \longrightarrow & G \times_X Y & \xrightarrow{a} & Y & \longrightarrow & 1
\end{array}$$

$$
\begin{array}{lccccccccc}
(3) & 0 & \longrightarrow & M & \longrightarrow & Y \times_X Y & \xrightarrow{p_1} & Y & \longrightarrow & 1 \\
 & & & \| & & \downarrow{\scriptstyle p_2} & & \downarrow & & \\
(4) & 0 & \longrightarrow & M & \longrightarrow & Y & \longrightarrow & X & \longrightarrow & 1.
\end{array}
$$

Both sequences (2) and (3) split, the first because (1) is split and the second by the diagonal $Y \longrightarrow Y \times_X Y$. It is a familiar fact in extension theory (and reappears as IV.(3.6) in this formulation) that any two split sequences are equivalent, which means that $G \times_X Y \xrightarrow{(a,p_2)} Y \times_X Y$ is an equivalence. It can be seen directly (e.g. use the metatheorem) that a determines an action, evidently principal, of G on Y. Note, of course, that fibred product over X is precisely cartesian product in $\underline{Y}$.

(4.4) Considering the same diagram, we see that $(a,p_2): G \times_X Y \longrightarrow Y \times_X Y$ gives that $G \times_X Y$ and $Y \times_X Y$ are extensions of Y with the same kernel M, which implies that G and Y are extensions of X with the same kernel M, the first being split. Hence we have shown:

(4.5) <u>Theorem</u>. Let X be a group, M an X-module, G the split extension of X with kernel M. Then singular extensions of X with kernel M are equivalent to principal G-objects in $(\underline{Gp},X)$. Equivalent extensions correspond to isomorphic objects of $\underline{PLO}(G)$.

Proof. We have shown everything but the last, but that is obvious.

(4.6) <u>Proposition</u>. Let M,X,G be as above. Then

$$\mathrm{Der}(X,G) \cong (\underline{Gp},X)(X \longrightarrow X, G \longrightarrow X).$$

Proof. Note that the last is $\underline{Y}(1,G) = H^0(G)$. The proof is easy and also well-known. See the remark in the middle of p.255 of [4].

(4.7) Thus we have identified $H^0(G)$ with $H^0(X,M) = \mathrm{Der}(X,M)$ and

$H^1(G)$ with $H^1(X,M)$, the usual group of singular extensions of (2.2) corresponds, as far as it goes, with the usual one. It is also evident that the identical analysis would work for any of the standard equational categories: associative, commutative, Lie, Jordan rings or algebras, etc. In each of those categories, as well as any equational category in which there is a group law among the operations, each group object must be abelian.

(4.8) In all these categories of algebras we might consider a relative cohomology, relative to some suitable functor. In the common examples this functor is algebraic, i.e. induced by a map of triples, and hence exact. The most common is the underlying functor from a category of K-algebras of some type to K-modules. In that case the relative cohomology classifies, in dimension one, those singular extensions which are split as K-modules. The Hochschild cohomology of associative algebras is of this form, while the corresponding absolute cohomology was given by Shukla. See [BB] for some of the details and further references.

(4.9) The Baer sum of singular extensions is defined in the following way. Given

$$\begin{array}{ccccccc} & & & Y_1 & & & \\ & & \nearrow & & \searrow & & \\ 0 \longrightarrow M & & & & & X \longrightarrow 1 & , \\ & & \searrow & & \nearrow & & \\ & & & Y_2 & & & \end{array}$$

two extensions with the same kernel, we first form $Y_1 \times_X Y_2$ and then observe that there are two embeddings $M \rightrightarrows Y_1 \times_X Y_2$. When these are rendered equal (or coequalized), the result is the Baer sum. We may indicate the process as

$$M \rightrightarrows Y_1 \times_X Y_2 \longrightarrow Y_1 \divideontimes Y_2 ,$$

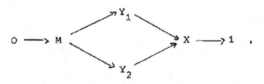

where $Y_1 \divideontimes Y_2$ is the Baer sum. In our generality, the embeddings $M \longrightarrow Y_i$ are replaced by actions $G \times Y_i \longrightarrow Y_i$, $i = 1,2$. The fibred product $Y_1 \times_X Y_2$ is simply the product in the category (<u>Gp</u>,X). Thus it seems more or less likely and is trivial to prove that the above sequence corresponds to our definition of the product in $H^1(G)$ (G commutative) given by the following diagram being a coequalizer:

$$Y_1 \times G \times Y_2 \rightrightarrows Y_1 \times Y_2 \longrightarrow Y_i \otimes_G Y_2.$$

This proves:

(4.10) <u>Theorem</u>: The equivalence between H'(G) and $H^1(X,M)$ given by (3.5) takes the tensor product multiplication in the first to the Baer sum in the second. Analogous results hold in the relative case.

Appendix: Giraud's theorem.

(A.1) After the completion of the five preceding chapters, I received from Ira Wolf a sketch of his proof of the Giraud theorem characterizing toposes. As I read it I realized that exact categories made a very convenient setting for the proof. This appendix presents a proof given along these lines. The proof is actually much closer to the one published by Verdier [Ve] than to Wolf's. It differs from the former in that it treats the question entirely in terms of Grothendieck topologies (in the sense of Artin) and that it involves neither a change of universe nor any essential use of an illegitimate category.

(A.2) The following terminology will be used throughout.
Let $\underline{C}$ be a category, C an object, $F\colon \underline{C}^{op} \longrightarrow \underline{S}$ a functor. A family of maps to C, $\{C_i \longrightarrow C\}$, is called a sieve (or a sieve on C). A sieve is called an F-sieve if every $C_i \times_C C_j$ exists and

$$FC \longrightarrow \Pi FC_i \rightrightarrows \Pi F(C_i \times_C C_j)$$

is an equalizer. It is called a universal F-sieve if for $C' \longrightarrow C$, every $C' \times_C C_i$ exists and $\{C' \times_C C_i \longrightarrow C'\}$ is an F-sieve. It is evident that if it is a universal F-sieve, then $\{C' \times_C C_i \longrightarrow C'\}$ will be universal also. If C'' is an object of $\underline{C}$, a sieve is called a (universal) C''-sieve if it is a (universal) $(-,C'')$-sieve. It is called a regular epimorphic sieve if it is a C''-sieve for every object C'' of $\underline{C}$ (this is an evident generalization of $\longrightarrow\!\!\!\!\rightarrow$) and a universal regular epimorphic sieve if it is a universal C''-sieve for every C'' of $\underline{C}$. These last two notions will be abbreviated r.e.s. and u.r.e.s. respectively.

(A.3) **Proposition**. Let $\{C_i \longrightarrow C\}$, and for each i, $\{C_{ij} \longrightarrow C_i\}$ be universal F-sieves. Then $\{C_{ij} \longrightarrow C\}$ is one also.

Proof. It is sufficient to show it is an F-sieve, since pullback commutes with composition. In order to do this we need the following lemma.

(A.4) Lemma. Let the diagram

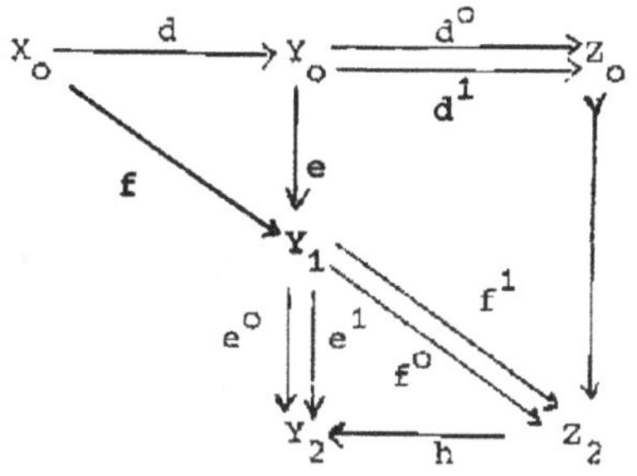

commute (that is, with d^o, e^o, f^o and with d^1, e^1, f^1), g be a monomorphism and e be the equalizer of e^o and e^1. Then d is the equalizer of d^o and d^1 if and only if f is the equalizer of f^o and f^1.

Proof. Chase the diagram.

(A.5) Now we return to the proof of (1.3). Apply the lemma with

$X_o = FC$, $Y_o = \prod_i FC_i$,

$Z_o = \prod_{i,k} F(C_i \times_C C_k)$, $Y_1 = \prod_{i,j} FC_{ij}$,

$Y_2 = \prod_{i,j,\ell} F(C_{ij} \times_{C_i} C_{i\ell})$, $Z_2 = \prod_{i,j,k,\ell} F(C_{ij} \times_C C_{k\ell})$.

The maps e and d are equalizers by assumption and we need only define h and show g is a monomorphism. The former is easily done by product projections. As for the latter, we define $Z_1 = \prod_{i,j,k} F(C_{ij} \times_C C_k)$. Now $\{C_{ij} \longrightarrow C_i\}$ is a universal F-sieve, so that by pulling back along the projection $\{C_i \times_C C_k \longrightarrow C_i\}$ we find that $\{C_{ij} \times_C C_k \longrightarrow C_i \times_C C_k\}$ is an F-sieve. This implies at least that $F(C_i \times_C C_k) \rightarrowtail \prod_j F(C_{ij} \times_C C_k)$ or that

$$\prod_{i,k} F(C_i \times_C C_k) \rightarrowtail \prod_{i,j,k} F(C_{ij} \times_C C_k),$$

which is $Z_o \rightarrowtail Z_1$. Similarly, $\{C_{kl} \longrightarrow C_k\}$ is a universal F-sieve, and by pulling it back along $C_{ij} \times_C C_k \longrightarrow C_k$ we see that $\{C_{ij} \times_C C_{kl} \longrightarrow C_{ij} \times_C C_k\}$ is an F-sieve too. Thus

$$F(C_{ij} \times_C C_k) \rightarrowtail \prod_{l} F(C_{ij} \times_C C_{kl}),$$

and by taking products over i,j,k we find $Z_1 \rightarrowtail Z_2$.

(A.6) <u>Proposition</u>. If $\{C_i \longrightarrow C\}$, and for each $\{C_{ij} \longrightarrow C_i\}$ are u.r.e.s, then so is $\{C_{ij} \longrightarrow C\}$.

(A.7) From the previous proposition it is clear that the class of all u.r.e.s. in a category $\underline{C}$ forms a topology, called the canonical topology. Any topology less fine than the canonical topology is called a standard topology.

(A.8) Another consequence of this proposition is that the usual assumption in a Grothendieck topology that the composition of covers is a cover (I.(4.1).b) is unnecessary. In fact, it is an easy corollary that given an arbitrary collection of sieves, the sheaves for the coarsest topology it generates are exactly those F for which every one of the given sieves is a universal F-sieve.

(A.9) <u>Proposition</u>. Let $\underline{C}$ have pullbacks. Then a topology on $\underline{C}$ is a standard topology if and only if every representable functor is a sheaf.

The proof is very easy and is omitted.

(A.10) Let $\underline{E}$ be a category. $\underline{E}$ is called a topos if

a) $\underline{E}$ has finite limits.

b) $\underline{E}$ has disjoint universal sums.

c) $\underline{E}$ is exact.

d) $\underline{E}$ has a set of generators.

The precise meanings of these follow. a) is clear. b) means that for every family $\{E_i\}$ of objects there is a sum $\coprod E_i$; that the square

$$\begin{array}{ccc} \delta_{ij}E_i & \longrightarrow & E_i \\ \downarrow & & \downarrow \\ E_j & \longrightarrow & \coprod E_i \end{array}$$

is a pullback where

$$\delta_{ij}\ E_i = \begin{cases} E_i & \text{if } i = j \\ 0, & \text{the initial object, when } i \neq j; \end{cases}$$

and that given $E_i \longrightarrow E \longleftarrow E'$, $E' \times_E \coprod E_i \cong \coprod (E' \times_E E_i)$ by the natural map. By interpreting this condition when $i \in \emptyset$, we see that $E' \times_E 0 = 0$ for any $E' \longrightarrow E$ and if $E' \longrightarrow 0$, that $E' \cong E' \times_0 0 \cong 0$. This implies that 0 is empty and will henceforth be denoted by $\emptyset$. c) is used in the sense of this paper and d) in the sense of II.(1.3); that is, there is a set Γ of objects such that for any $E \rightarrowtail E'$ not an isomorphism there is a $G \in \Gamma$ and a map $G \longrightarrow E'$ which does not factor through E.

(A.11) Theorem (Giraud). Let $\underline{E}$ be a category. Then the following are equivalent.

a) There is a small category $\underline{C}$ with finite limits such that $\underline{E} = \mathfrak{F}(\underline{C}^{op},\underline{S})$ for the canonical topology on $\underline{C}$.

b) There is a small category $\underline{C}$ such that $\underline{E} = \mathfrak{F}(\underline{C}^{op},\underline{S})$, sheaves for some topology on $\underline{C}$.

c) There is a small category $\underline{C}$ and a full embedding $I: \underline{E} \longrightarrow (\underline{C}^{op},\underline{S})$ which has an exact left adjoint.

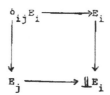

d) $\underline{E}$ is a topos.

e) $\underline{E} = \mathfrak{F}(\underline{E}^{op},\underline{S})$,(canonical topology) and has a set of generators.

(A.12) It is obvious that a) $\Longrightarrow$ b). That b) $\Longrightarrow$ c) is found in [Ar] and since the setting of exact categories in no way improves his proof, we omit it. The only thing to note in this connection is that if $P \rightarrowtail F$ where $P,F\colon \underline{C}^{op} \longrightarrow \underline{S}$ and F is a sheaf (in some topology), then the sheaf P^* associated to P is the subfunctor of F gotten by adding to PC every point in $FC \cap \Pi FC_i$ where $\{C_i \longrightarrow C\}$ is a cover in the topology. This obviously works even when $\underline{C}$ is large and the associated sheaf functor may not exist. The P^* so constructed can easily be seen to have the required universal mapping property: $(P^*,F) \cong (P,F)$ when F is a sheaf.

(A.13) Proposition. Condition c) $\Longrightarrow$ condition d).

Proof. Suppose $I\colon \underline{E} \longrightarrow (\underline{C}^{op},\underline{S})$ is a full embedding with left adjoint J. Then sums (as well as other colimits) are computed in $\underline{E}$ by $\coprod E_i = J\coprod IE_i$. We leave to the reader the easy task of showing that $(\underline{C}^{op},\underline{S})$ is itself a topos. In what follows we automatically identily the composite JI with the identity functor on $\underline{E}$. Then for a family $\{E_i\}$ of objects of $\underline{E}$.

$$\begin{array}{ccc} \delta_{ij}IE_i & \longrightarrow & IE_i \\ \downarrow & & \downarrow \\ IE_j & \longrightarrow & \coprod IE_i \end{array}$$

is a pullback. If we apply J and recall that J preserves initial objects, we get that

$$\begin{array}{ccc} \delta_{ij}E_i & \longrightarrow & E_i \\ \downarrow & & \downarrow \\ E_j & \longrightarrow & \coprod E_i \end{array}$$

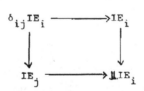

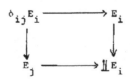

is a pullback. Similarly, given $E_i \longrightarrow E$ and $E' \longrightarrow E$, we have

$$E' \times_E \coprod E_i \cong JIE' \times_{JIE} J(\coprod IE_i)$$

$$\cong J(IE' \times_{IE} \coprod IE_i) \cong J(\coprod IE' \times_{IE} IE_i)$$

$$\cong J(\coprod I(E' \times_E E_i)) \cong \coprod (E' \times_E E_i).$$

Thus $\underline{E}$ has universal disjoint sums. If

$$\begin{array}{ccc} E'_o & \longrightarrow & E'_1 \\ \downarrow & & \downarrow \\ E_o & \twoheadrightarrow & E_1 \end{array}$$

is a pullback in $\underline{E}$, apply I and factor $IE_o \longrightarrow IE_1$ to get

$$\begin{array}{ccccc} IE'_o & \twoheadrightarrow & F' & \rightarrowtail & IE'_1 \\ \downarrow & & \downarrow & & \downarrow \\ IE_o & \twoheadrightarrow & F & \rightarrowtail & IE_1 \end{array} .$$

F' is defined to make the right hand square a pullback, and since the whole square is a pullback, so is the left hand square, whence $IE'_o \twoheadrightarrow F'$ as shown. The functor J preserves both $\twoheadrightarrow$ and $\rightarrowtail$ (the latter because it preserves finite limits), so we can apply I to get

$$\begin{array}{ccccc} E'_o & \twoheadrightarrow & JF' & \rightarrowtail & E'_1 \\ \downarrow & & \downarrow & & \downarrow \\ E_o & \twoheadrightarrow & JF & \rightarrowtail & E_1 \end{array} .$$

in which both squares are pullbacks. But since $E_o \twoheadrightarrow E_1$, it follows that $JF \twoheadrightarrow E_1$, whence $JF \xrightarrow{\sim} E_1$, and then $JF' \xrightarrow{\sim} E'_1$, which implies that $E'_o \twoheadrightarrow E'_1$. Thus the pullback of a regular epimorphism is also a regular epimorphism.

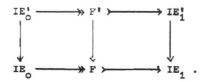

Suppose $E_1 \rightrightarrows E_o$ is an equivalence on E_o. It is clear from I(5.3) that a limit preserving functor preserves equivalence relations, so that there is an exact sequence

$$IE_1 \rightrightarrows IE_o \longrightarrow F$$

in $(\underline{C}^{op},\underline{S})$, and since J is exact,

$$E_1 \rightrightarrows E_o \longrightarrow JF$$

is an exact sequence as well. Thus $\underline{E}$ is exact.

Finally, if $E \rightarrowtail E'$ is not an isomorphism, it follows, since I is full and limit preserving, that $IE \rightarrowtail IE'$ and is not an isomorphism. This means there is a $C \in \underline{C}$ with $IEC \rightarrowtail IE'C$ not an isomorphism or, by the Yoneda lemma, a map $(-,C) \longrightarrow IE'$ which does not factor through IE. In view of adjointness, this is the same as a map $J(-,C) \longrightarrow E'$ which does not factor through E. Thus the objects $J(-,C)$, $C \in \underline{C}$ generate $\underline{E}$.

This completes the proof of (A.13).

(A.14) Now we turn our attention to showing d) $\Longrightarrow$ e). Until that is finished, $\underline{E}$ denotes a topos; $\mathfrak{F}(\underline{E}^{op},\underline{S})$, the category of sheaves in the canonical topology; and $R: \underline{E} \longrightarrow \mathfrak{F}(\underline{E}^{op},\underline{S})$, the embedding as representable functors.

(A.15) Proposition. R is exact.

Proof. The proof of I(4.3) is equally valid for any topology less fine than the canonical and finer than the regular epimorphism topology.

(A.16) Proposition. Let F be a sheaf. Then $F(\coprod E_i) = \Pi F E_i$ for any family of objects E_i of $\underline{E}$.

Proof. First observe that $\{E_i \longrightarrow \emptyset\}_{i\in\emptyset}$ is a cover. This is so since for any E",

$$= (\emptyset,E") \longrightarrow \prod_{i\in\emptyset} (E_i,E") \rightrightarrows \prod_{i,j\in\emptyset} E_i \times_{\emptyset} E_j,E")$$

is an equalizer, while there are no non-trivial $E' \longrightarrow \emptyset$ to pull back along. Replacing $(-,E")$ by any sheaf F, we see that $F\emptyset = 1$. Now let $E = \coprod E_i$. Since $E_i \times_E E_j = \delta_{ij}E_i$ we have, for any E", that

$$(E,E") \longrightarrow \Pi(E_i,E") \rightrightarrows \Pi(E_i \times_E E_j,E")$$

is an equalizer (all maps being isomorphisms). Hence $\{E_i \longrightarrow E\}_{i\in\emptyset}$ is an r.e.s. and, using the universality of the sums, it is easily seen to be a u.r.e.s. Then for any sheaf F,

$$FE \longrightarrow \Pi FE_i \rightrightarrows \Pi F(E_i \times_E E_j)$$

is an equalizer. Since $E_i \times_E E_j = \delta_{ij}E_i$ and $F\emptyset = 1$, the third term is the same as the second, which implies that $FE = \Pi FE_i$.

(A.17) <u>Proposition</u>. R preserves sums.

Proof. For any F and any $\{E_i\}$, $(R\coprod E_i,F) = F(\coprod E_i) = \Pi FE_i = \Pi(RE_i,F)$ $= (\coprod RE_i,F)$.

(A.18) <u>Proposition</u>. Every map of $\mathfrak{F}(\underline{E}^{op},\underline{S})$ factors as $\cdot \twoheadrightarrow \cdot \rightarrowtail \cdot$.

Proof. Let $F' \longrightarrow F$ be a map. Let P be the image as a functor. Then

$$F' \times_F F' \rightrightarrows F' \twoheadrightarrow P$$

is an exact sequence of functors and $F' \times_F F'$ is a sheaf. Since $P \rightarrowtail F$, P has an associated sheaf $P^* \rightarrowtail F$, which satisfies the universal mapping property that for F" a sheaf, $(P,F") = (P^*,F")$. From this, we see that $F' \times_F F' \rightrightarrows F' \longrightarrow P^*$ is exact in $\mathfrak{F}(\underline{E}^{op},\underline{S})$ while $P^* \rightarrowtail F$ (see (A.12)).

(A.19) <u>Proposition</u>. A sieve $\{E_i \longrightarrow E\}$ is a cover in the canonical topology if and only if $\coprod E_i \twoheadrightarrow E$.

Proof. The "only if" is trivial. Suppose $\coprod E_i \twoheadrightarrow E$. Then

$$(\coprod E_i) \times_E (\coprod E_i) \rightrightarrows \coprod E_i \longrightarrow E$$

is exact. The kernel pair is

$$\coprod E_i \times_E \coprod E_i \cong \coprod(E_i \times_E \coprod E_j) = \coprod(E_i \times_E E_j),$$

so that

$$\coprod(E_i \times_E E_j) \rightrightarrows E_i \longrightarrow E$$

is exact, from which

$$(E,E') \longrightarrow \Pi(E_i,E') \rightrightarrows \Pi(E_i \times_E E_j,E')$$

is an equalizer for all E and $\{E_i \longrightarrow E\}$ is an r.e.s. The universality follows easily from that of sums.

(A.20) Proposition. The set of objects RG, with $G \in \Gamma$, is a set of generators for $\mathfrak{F}(\underline{E}^{op},\underline{S})$.

Proof. Suppose $F \rightarrowtail F'$ is a monomorphism of sheaves such that $FG \xrightarrow{\sim} F'G$ for each $G \in \Gamma$. We will show that $F \xrightarrow{\sim} F'$. Let B be an object and find $\coprod G_i \twoheadrightarrow E$ with each $G_i \in \Gamma$. Then $\{G_i \longrightarrow E\}$ is a cover and hence we have the commutative diagram

$$\begin{array}{ccccc} FE & \longrightarrow & \Pi FG_i & \rightrightarrows & \Pi F(G_i \times_E G_j) \\ \downarrow & & \downarrow \cong & & \downarrow \\ F'E & \longrightarrow & \Pi F'G_i & \rightrightarrows & \Pi F'(G_i \times_E G_j), \end{array}$$

whose rows are equalizers, and an easy diagram chase shows $FE \xrightarrow{\sim} F'E$.

(A.21) Proposition. For any sheaf F, there is a regular epimorphism $RE \twoheadrightarrow F$.

Proof. Since $\mathfrak{F}(\underline{E}^{op},\underline{S})$ has $\cdot \twoheadrightarrow \cdot \rightarrowtail \cdot$ factorizations, we can repeat the argument of II(1.4) to see that

$$R(\coprod_{G \in \Gamma} \coprod_{(RG,F)} G) = \coprod\coprod RG \twoheadrightarrow F.$$

Proposition. Every sheaf is representable.

Proof. Consider the sequence

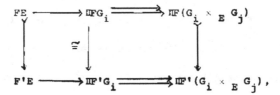

$$F' \rightrightarrows RE \longrightarrow F$$

where $RE \twoheadrightarrow F$ and F' is the kernel pair. Again we can find $RE' \twoheadrightarrow F'$.

Now we have $E' \longrightarrow E \times E$, which factors $E' \twoheadrightarrow E'' \rightarrowtail E \times E$, and since R is exact,

$$RE' \twoheadrightarrow RE'' \rightarrowtail R(E \times E),$$

and by the uniqueness of the factorization, $RE'' \cong F$. Then

$RE'' \rightrightarrows RE$ is an

equivalence relation and R is a full exact embedding, so that $E'' \rightrightarrows E$ is one too. Then there is an exact sequence

$$E'' \rightrightarrows E \longrightarrow E''',$$

and again, since R is exact, $RE''' \cong F$.

This completes the proof that d) $\Longrightarrow$ e).

(A.22) From now on $\underline{E}$ will be a category in which every sheaf for the canonical topology is representable. We suppose that $\underline{C}$ is a subcategory of $\underline{E}$ which is closed under subobjects and finite products and which contains a set of generators. Note that every sheaf's being representable implies that $\underline{E}$ has all limits. Our aim is to show that $\underline{E} \cong \mathfrak{F}(\underline{C}^{op},\underline{S})$ for the canonical topology on $\underline{C}$.

We say that a sieve $\{E_i \longrightarrow E\}$ is an extremal sieve if there is no subobject of E which factors each of the maps.

(A.23) Proposition. A sieve in $\underline{E}$ is extremal if and only if it is a cover in the canonical topology.

Proof. The "if" part is easy. For if $E' \rightarrowtail E$ were a subobject factoring all the $E_i \longrightarrow E$, then the fact that $(-,E')$ is a sheaf would provide an inverse to the inclusion $E' \rightarrowtail E$. To go the other way, suppose a sieve is extremal. Let $P: \underline{E}^{op} \longrightarrow \underline{S}$ be defined by

$PE_1 = \{f\colon E_1 \longrightarrow E \mid f$ factors through at least one $E_i \longrightarrow E\}$. Then $P \rightarrowtail (-,E)$, and by the remark (A.12) there is a sheaf $P^* \rightarrowtail (-,E)$ associated to P. If $P^* = (-,E')$, then $E' \rightarrowtail E$ factors every $E_i \longrightarrow E$, so $P^* = (-,E)$. Now in the category $(\underline{E}^{op},\underline{S})$,

$$\coprod(-,E_i) \times_P \coprod(-,E_i) \rightrightarrows \coprod(-,E_i) \longrightarrow P$$

is exact. Since $P \rightarrowtail (-,E)$, we have

$$\coprod(-,E_i) \times_P \coprod(-,E_i) \cong \coprod(-,E_i) \times_{(-,E)} \coprod(-,E_i)$$
$$\cong \coprod[(-,E_i) \times_{(-,E)} (-,E_j)] \cong \coprod(-,E_i \times_E E_j),$$

so that

$$\coprod(-,E_i \times_E E_j) \rightrightarrows \coprod(-,E_i) \longrightarrow P$$

is exact. Let E" be an arbitrary object. Then using the fact $(P,(-,E")) = (P^*,(-,E")) = (E,E")$ we hom this sequence into E" and have that

$$(E,E") \longrightarrow \Pi(E_i,E") \rightrightarrows \Pi(E_i \times_E E_j,E")$$

is an equalizer. Hence $\{E_i \longrightarrow E\}$ is an r.e.s. To show the universality, it is sufficient to show that for any $E' \longrightarrow E$, the sheaf associated to $P' = P \times_E (-,E')$ is $(-,E')$ itself. This is easily done by using the remark of (A.18) together with the usual proof that the associated sheaf functor is exact.*

(A.24) <u>Corollary</u>. The topology induced on $\underline{C}$ by the inclusion $\underline{C} \longrightarrow \underline{E}$ is the canonical topology.

Proof. Since $\underline{C}$ is closed under subobjects, a sieve $\{C_i \longrightarrow C\}$ is extremal in $\underline{C}$ if and only if it is in $\underline{E}$.

(A.25) This implies that there is a functor $I\colon \underline{E} \longrightarrow \mathcal{F}(\underline{C}^{op},\underline{S})$. This

---

* I am indebted to H. Schubert for pointing out an error in my original proof of this proposition.

functor is faithful, since $\underline{C}$ contains a set of generators of $\underline{E}$. If we can find a $J: \mathfrak{F}(\underline{C}^{op},\underline{S}) \longrightarrow \underline{E}$ such that JI = identity, it follows that I is an equivalence. Let $F: \underline{C}^{op} \longrightarrow \underline{S}$ be a sheaf. We extend it to a functor $\bar{F}: \underline{E}^{op} \longrightarrow \underline{S}$ in what by (A.23) is the only possible way. For $E \in \underline{E}$, choose an extremal sieve

$$\{C_i \longrightarrow E\},\ C_i \in \underline{C},$$

which certainly exists, since $\underline{C}$ contains a set of generators. Now let $\bar{F}E$ be defined so that

$$\bar{F}E \longrightarrow \Pi FC_i \rightrightarrows \Pi F(C_i \times_E C_j)$$

is an equalizer. Note that $C_i \times_E C_j \subset C_i \times C_j$ and hence is an object of $\underline{C}$ for all i,j. There remain two problems: to show that $\bar{F}$ doesn't depend on the choice of an extremal sieve and that it is a sheaf. First we need:

(A.26) Lemma. Let the diagram

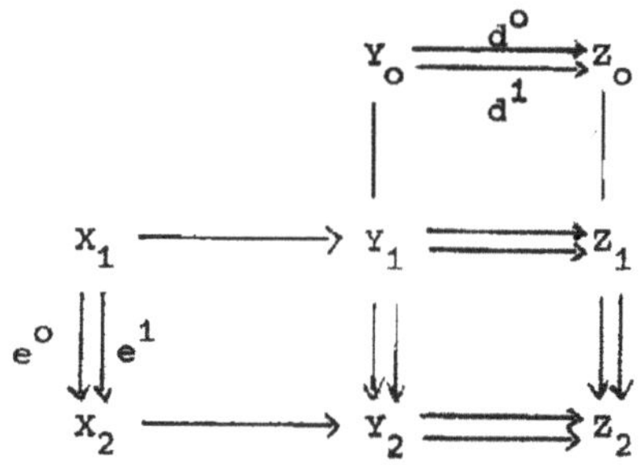

be commutative and the rows and columns be equalizers. Then the equalizer of d^o and d^1 is the same as that of e^o and e^1.

Proof. Chase the diagram.

(A.27) Proposition. $\bar{F}$ is well defined.

Proof. Let $\{C_i \longrightarrow E\}$ and $\{C'_k \longrightarrow E\}$ be two extremal sieves with

C_i, $C'_k \in \underline{C}$. Apply the above lemma with $Y_o = \Pi F C_i$, $Z_o = \Pi F(C_i \times_E C_j)$,

$X_1 = \Pi C'_k$, $X_2 = \Pi F(C'_k \times_E C'_\ell)$,

$Y_1 = \Pi F(C_i \times_E C'_k)$, $Z_1 = \Pi F(C_i \times_E C_j \times_E C'_k)$,

$Y_2 = \Pi F(C_i \times_E C'_k \times_E C'_\ell)$,

$Z_2 = \Pi F(C_i \times_E C_j \times_E C'_k \times_E C'_\ell)$. In all cases the products are taken over all available sets of indices.

(A.28) Proposition. $\bar{F}$ is a sheaf.

Proof. Let $\{E_i \longrightarrow E\}$ be an extremal sieve, and for each i, choose $\{C_{ij} \longrightarrow E_i\}$ an extremal sieve. Then $\{C_{ij} \longrightarrow E\}$ is an extremal sieve and can be used to define $\bar{F}E$. We now apply (A.4) with $X_o = \bar{F}E$, $Y_o = \Pi \bar{F} E_i$, $Z_o = \Pi \bar{F}(E_i \times_E E_\ell)$, $Y_1 = \Pi F(C_{ij})$, $Y_2 = \Pi F(C_{ij} \times_{E_i} C_{i\ell})$, $Z_2 = \Pi F(C_{ij} \times_E C_{k\ell})$. In applying the theorem in this direction, you do don't actually need g to be $\rightarrowtail$ if you know that $e^o.e = e^1.e$.

Thus $\bar{F}$ is a sheaf, and it is clear that $\bar{F}$ restricted to $\underline{C}$ is F. This completes the proof of Giraud's theorem.

References

[Ar] M. Artin, Grothendieck Topologies, Cambridge: Harvard University Press, 1962.

[B-C] M.Barr, Non-abelian full embedding, Outline, to appear in"Actes du Congrès International des Mathématiciens, 1970"

[B-M] M.Barr, Non-abelian full embedding,announcement of Results, in "Reports of the Midwest Category Seminar V", Lecture Notes in Math. 195, Berlin - Heidelberg - New York: Springer,1971.

[BB] M.Barr and J.Beck, Homology and standard constructions, in "Seminar on Triples and Categorical Homology Theory", Lecture Notes in Math. 80 Berlin - Heidelberg - New York: Springer,1969.

[Be] J.Beck, Triples, Algebras and Cohomology, Dissertation, Columbia University, 1967.

[BP] H.-B.Brinkmann and D.Puppe, Abelsche und exakte Kategorien, Korrespondenzen, Lecture Notes in Math. 96, Berlin - Heidelberg - New York: Springer, 1969.

[CE] H.Cartan and S.Eilenberg, "Homological Algebra", Princeton: Princeton University Press, 1958

[Ch] S.U. Chase, "Galois objects in Hopf Algebras and Galois Theory", Lecture Notes in Math. 97, Berlin - Heidelberg - New York: Springer, 1969.

[DP] A.Dold and D.Puppe, Homologie nicht additiver Funktoren, Anwendungen, Ann. Inst. Fourier 11, 201-312 (1961).

[Ge] M.Gerstenhaber, A categorical sitting for the Baer extension theory, in "Applications of Categorical Algebra", Proceedings of Symposia in Pure Mathematics, 17, Providence, Amer. Math. Soc. 50-64 (1970).

[Ke] G.M.Kelly, Monomorphisms, epimorphisms and pullbacks, J. Australian Math. Soc. 9, 124-142 (1969).

[LU] S.Lubkin, Imbedding of abelian categories, Trans. Amer. Math. Soc. 97, 410-417 (1960).

[Mac] S. Mac Lane, "Homology", Berlin - Heidelberg - New York: Springer 1963.

[Man] E. Manes, "A Triple Miscellany: Some Aspects of the Theory of Algebras over a Triple", dissertation, Wesleyan University, 1967.

[Mi] B. Mitchell, "Theory of Categories", New York and London Academic Press, 1965.

[Ul] F.Ulmer, Locally α-presentable and locally α-generated categories, (Summary of joint work with P.Gabriel), in "Reports of the Midwest Category Seminar V," Lecture Notes in Math. 195, Berlin - Heidelberg - New York: Springer 1971.

[Ve] J-L. Verdier, Topologies et faisceaux, S.G.A.A., fasc. I, 1963-1964.

REGULAR CATEGORIES

Pierre Antoine Grillet

INTRODUCTION

Decompositions of morphisms into mono- and epimorphisms occur in nearly all the examples which justify the very existence of category theory. Thus it is not surprising that they received attention very early, with the emergence of abelian categories and, in the non-abelian case, MacLane's 1948 paper. It seems much more surprising that further developments has to await more than a decade for the work of Isbell and Barr and others, and also that satisfactory ways to describe non-abelian algebraic phenomena (triples, monoids etc.) did not appear until about the same time, and do not use decompositions. It would seem that, in non-abelian situations, the apparent lack of good properties may have made the actual manipulation of mono-epimorphism decompositions seem unable to attain enough versatility to be of any use in proving things, so that other methods had to be devised.

All the same, decompositions are there, and as categories are expected to account for more and more phenomena it becomes more and more difficult and unnatural not to use them. This may be the basic reason why in the last decade more and more people have been talking decompositions, each time in a slightly different form, but with similar ideas in mind. Also, it is not a denigration of triples and/or monoids to say that by their very nature they cannot by themselves always account for algebraic phenomena with the desired combination of generality and precision that is necessary in some situations (VanOsdol's contribution to this volume is a case in point).

As far as algebraic situations are concerned, the consideration of regular categories may fill these needs very neatly. A regular category (also considered in Micheal Barr's part with weaker but essentially similar axioms) is a finitely complete category in which every morphism f has a decomposition $f = mp$ where m is a monomorphism and p a regular epimorphism (= a coequalizer), and where pullbacks carry regular epimorphisms (i.e. if $fg' = gf'$ is a pullback and f is a regular epimorphism, then so is f'). There is considerable evidence that regular categories can play with regard to non-abelian algebra the role that abelian categories play in abelian algebra. Examples include varieties (finitary or infinitary) as well as abelian categories, and regularity transfers well to categories of functors, algebras over a triple and sheaves. Just as abelian categories can account for all elementary aspects of life with modules (kernels, hom groups, exact sequences etc), all elementary manipulations of subobjects and congruences that are possible in a variety are equally possible in any regular category. [This includes one more (slightly different) account of decompositions, subobjects and relations; but this time it seems that regular categories provide the right context for all this. Indeed all properties one would expect of a satisfactory account are obtained, and there is evidence that the axioms cannot be significantly weakened and still accomplish this.] The rest of the evidence is the behavior of sheaves in a a regular category, and the fact that they provide the adequate concept for generalizations of Mitchell's full embedding theorem.

This author's contribution is divided into three parts. The first part gives an account of decompositions and relations in a regular category, as well as the easier examples and transfer properties. In the second part are given necessary and sufficient conditions that directed colimits in a cocomplete regular category preserve monomorphisms and finite limits; directed colimits then show additional instances of good behavior. The last part deals with sheaves in suitable regular categories. More can be found in the introduction of each part.

All three parts have been written so that only a minimal knowledge of the bare essentials of category theory (a fraction of [31], and the definition of a triple) and universal algebra (available in [7], [32]) is necessary for the text. The notation and terminology are as in Mitchell [31] with the following exceptions. Diagrams are defined as functors from a small category. In order that the text make sense in everybody's set theory, in which there may not exist choice functions in classes, we have used the following conventions regarding existence statements: taking as example the existence of limits, if we merely wish to say that there exists a limit to every diagram in C, we say "C is with limits"; if we wish to say that there is a function which selects a limit for every diagram in C, we say "C has limits". Complete, cocomplete and well-powered are to be read as "has", not "is with". Of course this makes no difference if C is small; in general we have kept the selecting functions as inobstrusive as possible. Subobjects are defined as equivalence classes of monomorphisms (where the monomorphisms m and n are equivalent in case $m = ni$ for some isomorphism i). The equalizer Equ(f,g) is a subobject, and an element thereof is just an equalizer of f and g ; similar conventions apply to intersections and dually to quotient-objects, coequalizers and cointersections. We start from definitions of images and unions which differ from Mitchell's as indicated in the text.

One of the changes in notation is not a trifle. Products when used as functors are denoted by $\prod$ (π for finite products). Thus $\prod_{i\in I} f_i$ denotes the morphism $\prod_{i\in I} A_i \longrightarrow \prod_{i\in I} B_i$ induced by all $f_i : A_i \longrightarrow B_i$. To denote the morphism induced by all $f_i : A \longrightarrow B_i$, i.e. $A \longrightarrow \prod_{i\in I} B_i$, we use the notation $\mathop{\mathsf{X}}_{i\in I} f_i$ (x for finite products). This allows to denote coproducts by $\coprod$ and we think that the confusion it may create is less than that of having to contend with $(f_i)_{i\in I}$ instead of $\mathop{\mathsf{X}}_{i\in I} f_i$ in numerous proofs.

I. EXAMPLES AND ELEMENTARY PROPERTIES

This part is divided into six sections. Sections 1 and 2 contain definition and examples of regular categories. Decompositions of various kinds are investigated in section 1, paving the way for the definition of regular categories which begins section 2. In section 2, we also show that when $\mathcal{C}$ is a regular category and $\mathcal{I}$ is a small category, then the functor category $[\mathcal{I},\mathcal{C}]$ is regular (and the evaluation functor $[\mathcal{I},\mathcal{C}] \longrightarrow \mathcal{C}^{\mathcal{I}}$ preserves and reflects regular decompositions); a similar result is proved for $\mathcal{C}^{\mathbb{T}}$, when $\mathbb{T} = (T,\mu,\epsilon)$ is a triple on $\mathcal{C}$ such that T preserves regular epimorphisms.

Sections 3,4 and 5 concern the calculus of subobjects, relations and congruences respectively, in a regular category.

The last section gives various properties of limits and colimits in a regular category, as well as completeness → cocompleteness implications. A synopsis of the main formulae in the middle part will be found at the end of that section.

We have tried to make the exposition as careful as possible, especially in giving additional justifications for the definitions and ways of doing things. Relations, and congruences, are defined as subobjects of products, rather than pairs of morphisms, or kernel pairs. Factorization systems, on which the emphasis has been historically, are but briefly considered; the main reason, explained in more detail at the end of section 5, is that they would bring very little additional generality, and this, we think, is not justified by the examples. The one advantage of using factorization systems would be to explain the duality of sorts, which is very apparent throughout, between monomorphisms and regular epimorphisms, subobjects and congruences, etc.; however, no perfectly self-dual account can be given, because the duals of several important properties just do not hold in varieties.

All the results here have been announced in [14]. While they have not otherwise been published before under that form, there is little claim of originality that can be laid for the contents of sections 3-5, since these have been considered before, in part and under sundry guises, by a great many people (most notably, [26],[27], [19],[20],[28],[33],[34],[13],[24],[23],[5],[3],[1],[2],[9]); the part on congruences is certainly the least unoriginal: congruences have been considered before, e.g. as kernel pairs as in [25], but this does not allow for all the manipulations that are possible here, or at least not in a way which is both satisfactory and natural. Most of the references above have to do with factorization systems, which likewise takes care of section 1. Only the most glaring cases of overlap have been indicated in the text.

1. DECOMPOSITIONS.

1. Let $\mathbf{C}$ be any category and f be a morphism of $\mathbf{C}$. A decomposition of f (also known as a mono-epi decomposition, or factorization) is a pair (m,p) of a monomorphism m and epimorphism p such that $f = mp$; $\mathbf{C}$ is a category with decompositions in case this exists for every $f \in \mathbf{C}$.

In general a preorder (= reflexive and transitive relation) is defined on decompositions of a given $f \in \mathbf{C}$ as follows: if (m,p), (n,q) are decompositions of f, then $(m,p) \leq (n,q)$ if and only if there exists a morphism $u \in \mathbf{C}$ such that the following diagram

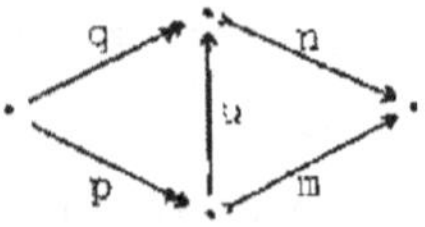

commutes. Note that u is necessarily a bimorphism. When there exists a diagram as above in which u is an isomorphism, the decompositions (m,p) and (n,q) are equivalent; this happens if and only if (m,p) $\leq$ (n,q) and (n,q) $\leq$ (m,p) .

Granted that f has decompositions, it is natural to look for decompositions of f which are maximal or minimal, or even greatest or least, under the preorder relation. Indeed the general decompositions are not good for much, and in nice categories every morphism has a decomposition with one or the other of these properties; for instance in an abelian category all the decompositions of a given morphism are equivalent, hence they are all greatest and least. This is a rather extreme situation, but in the category of sets, and in that of groupoids (= sets with one binary operation, which does not have to be nice), as well as in the category of all groups, every decomposition is also greatest and least. Anticipating a little, in a variety every morphism has a least decomposition (the obvious one); but it need not be greatest. In fact, in the variety of semigroups there are morphisms which do not have a greatest decomposition [22]. In other situations, such as in the category of all topological spaces, every continuous mapping $f : X \longrightarrow Y$ has a least decomposition $X \longrightarrow f(X) \longrightarrow Y$ (with the quotient topology on f(X)) and a greatest decomposition $X \longrightarrow f(X) \longrightarrow Y$ (with the subspace topology on f(X)) and they may be distinct; a similar result holds for Hausdorff topological spaces (with $\overline{f(X)}$ in the greatest decomposition), and also for partially ordered sets and order-preserving mappings, and for hypergroupoids (= sets with one multivalued binary operation); more topological examples can be found in [17]. In conclusion, in purely algebraic situations (sets with single-

-valued operations defined everywhere), we can expect least-decompositions (and should not expect the other kinds).

2. The classical approach to least decompositions is either to consider only certain decompositions and set forth axioms which among other things insure one of them will be least (that is using factorization systems); or to consider decompositions (m,p) with an additional condition on p. This is the way minimal decompositions arise: (m,p) is a minimal decomposition if and only if in every decomposition (n,q) of p, n is an isomorphism (then p is called extremal [20], [21]). Unfortunately, there is no known necessary and sufficient condition on p that (m,p) be a least decomposition for every m; but there are a great many sufficient conditions. That p be a retraction is one of them, although too strong to be generally useful in that context. Of more interest are the following:

p is <u>regular</u>, i.e. is a coequalizer (used in that sense in [25], [3] et al.);

p is <u>strict</u> (cf.[21]), i.e.(without the set-theoretical sophistication of [21]) a (small) cointersection of coequalizers;

p is <u>subregular</u> (called special in [18], fermé (closed) in [13]), i.e. whenever g has same domain as p, and $pu = pv$ implies $gu = gv$, then $g = tp$ for some (unique) t;

p is <u>strong</u> [2], i.e. $fp = mg$, with m a monomorphism, implies $g = tp$ for some (unique) t.

<u>Proposition</u> 1.1. Let p be an epimorphism. Each of the following conditions implies the next one: i) p is regular; ii) p is strict; iii) p is subregular; iv) p is strong; v) whenever (m,p) is a decomposition of f it is a least decomposition of f.

<u>Proof</u>. Trivially i) implies ii). To show that ii) implies

iii), let p be a cointersection of coequalizers $p_i \in \mathrm{Coequ}(a_i, b_i)$ $(i \in I)$, in particular p factors through every p_i $(p = t_i p_i)$. If $pu = pv$ implies $gu = gv$, then for every i we have $pa_i = t_i p_i a_i = t_i p_i b_i = pb_i$, so that $ga_i = gb_i$ and g factors through every p_i; hence it factors through their cointersection p, which shows that p is subregular. Next, assume that p is subregular and that $fp = mg$, where m is a monomorphism; then $pu = pv$ implies $mgu = fpu = fpv = mgv$ and $gu = gv$; hence $g = tp$ for some t. If finally p is strong, and (m,p), (n,q) are decompositions of f, then $nq = mp$ implies $q = tp$ for some t; then also $mp = ntp$, and $m = nt$; therefore $(m,p) \leq (n,q)$, which shows that (m,p) is a least decomposition of f.

A _decomposition_ is called _regular_ (_strict_, _subregular_, _strong_) when the epimorphism therein is regular (strict, subregular, strong). All are least decompositions, by 1.1. In the usual cases, all five concepts are equivalent, so that the initial choice of conditions is not of extreme importance; more precisely, we have the following results.

First, call $\mathbf{C}$ _regularly co-well-powered_ if for each object $A \in \mathbf{C}$ there exists a set $\mathfrak{C}$ of regular epimorphisms of domain A, such that every regular epimorphism p of domain A is equivalent to some $q \in \mathfrak{C}$ (more precisely, in accordance with our conventions, there is a choice function F, such that $q = F(p)$ always serves).

Proposition 1.2. If $\mathbf{C}$ has coequalizers and is regularly co-well-powered, then strict and subregular are equivalent.

Proof. Let p be a subregular epimorphism. Let $(q_i)_{i \in I}$ be the family of all $q_i \in \mathfrak{C}$ such that p factors through q_i; we shall prove that p is a cointersection of $(q_i)_{i \in I}$. Let g have same domain as p and factor through every q_i. If $pu = pv$, then there is some $q \in \mathfrak{C}$ with $q \in \mathrm{Coequ}(u,v)$; in fact, $q = q_i$ for some i since p factors

$q \in \mathrm{Coequ}(u,v)$; then $g = sq$ for some s and $gu = squ = sqv = gv$. Since p is subregular, it follows that $g = tp$ for some (unique) t . Thus p is a cointersection of regular epimorphisms, i.e. is strict. The converse is part of 1.1.

Recall that a <u>kernel pair</u> of a morphism f is any pair (x,y) such that $fx = fy$ is a pullback (any two such are equivalent in the obvious sense).

<u>Proposition</u> 1.3. In a category with kernel pairs, regular, strict and subregular are equivalent ; in fact an epimorphism satisfying any of these conditions coequalizes his kernel pair(s).

<u>Proof</u>. Let p be a subregular epimorphism and (x,y) be a kernel pair of p. Let g be such that $gx = gy$. If $pu = pv$, then $u = xs$, $v = ys$ for some s (since $px = py$ is a pullback) and therefore $gu = gv$. Then g factors (uniquely) through p, which proves that $p \in \mathrm{Coequ}(x,y)$, in particular p is regular. The remaining implications follow from 1.1.

<u>Proposition</u> 1.4. If C is with regular decompositions (or with kernel pairs and subregular decompositions), then regular, strict, subregular, strong and extremal are equivalent.

<u>Proof</u>. Let p be an extremal epimorphism. There is a decomposition (n,q) of p with q regular (subregular); since p is extremal, it is equivalent to q hence also regular (subregular). The remaining implications follow from 1.1, 1.3.

The hypotheses of 1.2, 1.3, 1.4 are satisfied in any variety (see below).

3. Another way to obtain least decompositions is to deduce their existence from completeness or cocompleteness properties of the category.

Proposition 1.5. Let C be a well-powered category with intersections and decompositions. For every morphism $f \in C$ there exists a least decomposition of f.

Proof. Since C is well-powered, there exists a set $(m_i, p_i)_{i \in I}$ of decompositions of f, such that each decomposition of f is equivalent to one of these (note that, when (m,p) is a decomposition of f, p is uniquely determined by m). Let m be an intersection of all m_i; since f factors through every m_i, we have $f = mp$ for some p. First we show that p is an epimorphism. Let (n,q) be a decomposition of p. Then (mn,q) is a decomposition of f; hence there exist an i and an isomorphism v with $mn = m_i v$, $p_i = vq$. We also have $m = m_i u_i$ for some u_i. If v' is the inverse of v, then $mnv'u_i = m_i u_i = m$, $m_i u_i nv' = mnv' = m_i$; hence $nv'u_i = 1$, $u_i nv' = 1$, and nv' is an isomorphism. Therefore n is an isomorphism and $p = nq$ is an epimorphism. Thus (m,p) is a decomposition of f. Now m factors through every m_i, and this implies that $(m,p) \leq (m_i, p_i)$ for all i; it follows that (m,p) is a least decomposition of f.

Similar but vastly more sophisticated results can be found in [21]. Our last result is due to Tierney (mentioned in [3]):

Proposition 1.6. Let C be a category with pullbacks and coequalizers of kernel pairs. Assume that pullbacks in C carry regular epimorphisms. Then every morphism of C has a regular decomposition.

Proof. Take $f \in C$; there exists a kernel pair (x,y) of f and $p \in \mathrm{Coequ}(x,y)$; since $fx = fy$, we have $f = mp$ for some m and it suffices to show that m is a monomorphism. There exist pullbacks $mu = mv$, $pu' = up'$, $pv' = vq'$, $p'q'' = q'p''$; juxtaposing yields a pullback $(mp)(u'q'') = (mp)(v'p'')$, i.e. $f(u'q'') = f(v'p'')$; since $fx = fy$ is also a pullback, $px = py$ implies $pu'q'' = pv'p''$. But then $up'q'' = vq'p'' = vp'q''$; the hypothesis implies that p',q',q'' are epimorphisms, and it follows that $u = v$. Hence m is a monomorphism: $ma = mb$ implies $a = ux$, $b = vx$ for some x and $a = b$. Thus (m,p) is a regular decomposition of f.

4. The categories to be considered later are all with regular decompositions, not just least decompositions; this will provide us with more general factorization properties. Subregular decompositions would do just as well, but in the cases we are interested in, 1.4 makes them coincide with regular decompositions. On the other hand, strong decompositions are not quite strong enough for our purposes (see the end of section 5).

There are two basic examples of categories with regular decompositions. First are abelian categories (more generally, exact categories in the sense of [31]): the decompositions there are regular, since every epimorphism is then conormal, hence regular.

The other example is provided by varieties (finitary or infinitary). In a variety, every [homo]morphism has an obvious injective-surjective decomposition. To see that these are regular decompositions, we recall the construction of the pullback on $f : A \longrightarrow C$ and $g : B \longrightarrow C$ in a variety. Let $D = \{(a,b) \in A \pi B \ ; \ f(a) = g(b)\}$; this is a subalgebra of $A \pi B$ and therefore lies in the variety. The maps $x : (a,b) \longmapsto a$, $y : (a,b) \longmapsto b$ are homomorphisms such that $fx = gy$, and in fact $fx = gy$ is a pullback. If now f is a surjective homomorphism and we let $g = f$, then in the above D is but the congruence $\ker f$ induced by f; if h is any homomorphism such that $hx = hy$, then for every $(a,b) \in D$ we have $h(a) = h(x(a,b)) = h(y(a,b)) = h(b)$, in other words $\ker f \subseteq \ker h$, and it follows from the induced homomorphism theorem that h factors uniquely through f. This shows that $f \in \mathrm{Coequ}(x,y)$, so that f is regular.

Thus a variety has regular decompositions. Furthermore, every extremal epimorphism in a variety has a decomposition (m,p) with p surjective, in which m must be an isomorphism; it follows that extremal epimorphisms are surjective and by 1.1 regular, extremal, etc. are all equivalent to surjective.

It should also be noted that varieties satisfy the hypothesis of 1.6. Indeed consider the general pullback as above and assume that f is surjective. For each $b \in B$, there exists $a \in A$ with $f(a) = g(b)$, i.e. $(a,b) \in D$: this shows that y is surjective, and that pullbacks in a variety carry regular epimorphisms. (The same is true in abelian categories.)

All these properties of varieties are still true in any class of universal algebras which admits products and subalgebras.

5. We conclude with a few trivial results showing that in a category with regular decompositions, the regular decompositions behave just as well as the injective-surjective decompositions in a variety, and regular epimorphisms just as well as surjective mappings. The existence of regular decompositions is assumed throughout.

<u>Proposition</u> 1.7. Any two regular decompositions of the same morphism are equivalent.

<u>Proof</u>. By 1.1, both are least decompositions of that morphism.

<u>Proposition</u> 1.8. A morphism f is an isomorphism if and only if it is both a monomorphism and a regular epimorphism.

<u>Proof</u>. If f is both a monomorphism and a regular epimorphism, then $(f,1)$ is a decomposition of the extremal epimorphism f and so f must be an isomorphism. The converse is clear.

<u>Proposition</u> 1.9. Let $fa = bg$ be a commutative square and

(m,p), (n,q) be regular decompositions of f and g. There is a unique morphism t such that the following diagram commutes:

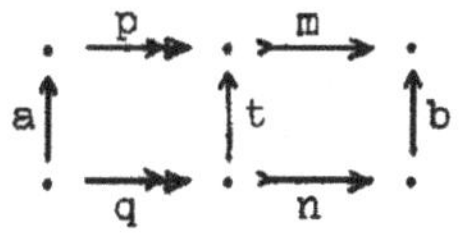

Proof. Since m is a monomorphism and q is strong, m(pa) = (bn)q implies pa = tq for some unique t. Then mtq = bnq shows that mt = bn as well. (One may call t "induced on the image")

Proposition 1.10. If f and g are regular epimorphisms and fg is defined, then fg is a regular epimorphism. [We are in a category with regular decompositions; the result is not true in general.]

Proof. Let (m,p) be a regular decomposition of fg. Since g is strong, mp = fg implies p = tg for some t; note that t is an epimorphism and that also f = mt (since fg = mtg). Hence (m,t) is a decomposition of f and since f is extremal this implies that m is an isomorphism. Hence fg is regular, like p.

Proposition 1.11. If fg is a regular epimorphism, then so is f.

Proof. Take regular decompositions (m,p) of f, (n,q) of g, and (k,r) of pn. Then (mk,rq) is a decomposition of fg, and since fg is (in particular) extremal, mk is an isomorphism. If v is the inverse isomorphism, then mkv = 1 shows that m is a retraction; but m is also a monomorphism, so that it is in fact an isomorphism. Hence f is regular, like p.

We have stated these results in the form we shall use later, but it is clear that 'strong' is the condition that makes them work (in fact they still hold if the category is only with strong decompositions and regular is replaced by strong everywhere). They imply that a category with regular decompositions ipso facto has a bifactorization system in the sense of [2], as well as a "bicategory" structure in the sense of [26],[27] (that is, if in fact the category *has* regular decompositions) and [24].

The last property is connected with products:

Proposition 1.12. Assume furthermore that the category has finite products and that pullbacks carry regular epimorphisms. Then every finite product of regular epimorphisms is a regular epimorphism.

Proof. It suffices to show that when f and g are regular epimorphisms then so is $f \sqcap g$. For this, we note that $f \sqcap g = (f \sqcap 1)(1 \sqcap g)$ and apply the hypothesis, 1.10 and the following

Lemma 1.13. Every diagram

$$\begin{array}{ccc} A \sqcap B & \longrightarrow & B \\ {\scriptstyle 1 \sqcap g}\downarrow & & \downarrow{\scriptstyle g} \\ A \sqcap B' & \longrightarrow & B' \end{array}$$

(where the horizontal maps are projections) is a pullback.

The proof of the lemma is left to the reader.

2. REGULAR CATEGORIES: DEFINITION AND EXAMPLES.

1. A *regular category* is a finitely complete category with regular decompositions, in which the following condition holds:

Pullback axiom: if $fg' = gf'$ is a pullback and f is a regular epimorphism, then f' is also a regular epimorphism.

Finite completeness implies that we could replace 'regular' by 'strict' or even by 'subregular' everywhere in the definition (by 1.3); in particular, all three conditions are equivalent in a regular category, and also are equivalent to 'strong' and to 'extremal' (by 1.4), although the last two would not give an equivalent definition. Finally, in a category with coequalizers, the existence of regular decompositions follows from the other axioms (1.6).

The two basic examples of regular categories are abelian categories and varieties of universal algebras (more generally, classes of universal algebras which admit products and subalgebras), as we have seen in the previous section. Of course the definition was calculated to include these examples. On the other hand the pullback axiom rules out the category of all topological spaces and similar examples (other than compact).

2. Additional examples of regular categories come from transfer theorems.

Theorem 2.1. Let $\mathfrak{X}$ be a small category and $\mathfrak{G}$ be a regular category. Then the functor category $[\mathfrak{X},\mathfrak{G}]$ is regular. Furthermore a morphism of $[\mathfrak{X},\mathfrak{G}]$ is a monomorphism (a regular epimorphism) if and only if it is a pointwise monomorphism (regular epimorphism).

Proof. Let η be a pointwise regular epimorphism of $\mathfrak{F} = [\mathfrak{X},\mathfrak{G}]$ and $\eta\alpha = \eta\beta$ be a pullback in $\mathfrak{F}$. For each $X \in \mathfrak{X}$, $\eta X.\alpha X = \eta X.\beta X$ is then a pullback in $\mathfrak{G}$ and since ηX is regular it follows (from 1.3) that $\eta X \in \mathrm{Coequ}_{\mathfrak{G}}(\alpha X,\beta X)$. Therefore $\eta \in \mathrm{Coequ}_{\mathfrak{F}}(\alpha,\beta)$ is a regular epimorphism. On the other hand a pointwise monomorphism is also a monomorphism.

If now η is an arbitrary morphism in $\mathfrak{F}$, we choose for each $X \in \mathfrak{X}$ a regular decomposition $(\mu X,\pi X)$ of ηX [this does not require that $\mathfrak{G}$ have regular decompositions since $\mathfrak{X}$ is small]. Put $\eta : F \longrightarrow G$ and let HX be the domain of μX. For each $f : X \longrightarrow Y$ we have a commutative diagram

$$\begin{array}{ccccc} FX & \xrightarrow{\pi X} & HX & \xrightarrow{\mu X} & GX \\ {\scriptstyle Ff}\downarrow & & & & \downarrow{\scriptstyle Gf} \\ FY & \xrightarrow{\pi Y} & HY & \xrightarrow{\mu Y} & GY \end{array}$$

and by 1.9 there is a unique morphism $Hf : HX \longrightarrow HY$ which keeps the diagram commutative. Because of the uniqueness it is then clear that we now have defined a functor H. In addition, the diagram shows that μ, π are natural transformations. By the first part of the proof, we have obtained a regular decomposition (μ, π) of η.

If in the above η is a monomorphism, then π is a monomorphism, hence an isomorphism by 1.8, so that η is a pointwise monomorphism (like μ). [This can also be proved using pullbacks.] If in the above η is a regular epimorphism, then so is μ by 1.11, so that μ is an isomorphism and η is a pointwise regular epimorphism (like π). It is then clear that the pullback axiom, as well as finite completeness, are inherited by $\mathfrak{F}$ from $\mathfrak{A}$, so that $\mathfrak{F}$ is a regular category, q.e.d.

The second part of the statement can again be expressed as follows. It follows from the theorem that the product category $\mathfrak{A}^{\mathrm{Ob}\mathcal{I}}$ (of all functors of the discrete category $\mathrm{Ob}\mathcal{I}$ into $\mathfrak{A}$) is a regular category, with pointwise regular decompositions. Then the evaluation functor $[\mathcal{I},\mathfrak{A}] \longrightarrow \mathfrak{A}^{\mathrm{Ob}\mathcal{I}}$ preserves and reflects regular decompositions.

Generally, a functor F between regular categories will be called <u>left exact</u> if it preserves finite limits (hence also monomorphisms) <u>right exact</u> if it preserves existing finite colimits (hence also regular epimorphisms), <u>exact</u> if it has both properties; <u>reflectively</u> left exact etc. are obtained by replacing "preserves" by "reflects" in the above. In particular, an exact (reflectively exact) functor preserves (reflects) regular decompositions. The terminology is close to Barr's [3], with slight modifications to fit abelian usage more closely (in spirit, at least). The evaluation functor in 2.1 is exact and reflectively exact. A similar result is true for algebras over a triple, except that right exactness cannot be expected:

Theorem 2.2. Let $\mathcal{G}$ be a regular category and $\mathbb{T} = (T, \eta, \mu)$ be a triple on $\mathcal{G}$ such that T preserves regular epimorphisms. Then $\mathcal{G}^{\mathbb{T}}$ is a regular category and the forgetful functor $\mathcal{G}^{\mathbb{T}} \longrightarrow \mathcal{G}$ preserves and reflects regular decompositions.

Proof. First, recall that the objects of $\mathcal{G}^{\mathbb{T}}$ are pairs (A,a) with $A \in \mathcal{G}$, $a : TA \longrightarrow A \in \mathcal{G}$, $a \cdot \eta A = 1$, $a \cdot \mu a = a \cdot Ta$; a morphism $f : (A,a) \longrightarrow (B,b) \in \mathcal{G}^{\mathbb{T}}$ is a morphism $f : A \longrightarrow B \in \mathcal{G}$ such that $fa = b \cdot Tf$ (see [8]). We already know that $\mathcal{G}^{\mathbb{T}}$ is as complete as $\mathcal{G}$, in particular is finitely complete ([30]; also easy to see directly).

The theorem itself is proved much as 2.1. Let $f \in \mathcal{G}^{\mathbb{T}}$. If f is a monomorphism in $\mathcal{G}$, then it is one in $\mathcal{G}^{\mathbb{T}}$. Now assume that f is a regular epimorphism in $\mathcal{G}$. Let $fx = fy$ be a pullback in $\mathcal{G}^{\mathbb{T}}$; this yields a pullback $fx = fy$ in $\mathcal{G}$ and since f is regular we have $f \in \mathrm{Coequ}_{\mathcal{G}}(x,y)$; we now show that this is still true in $\mathcal{G}^{\mathbb{T}}$. Put $f : (A,a) \longrightarrow (B,b)$ and let $g : (A,a) \longrightarrow (C,c) \in \mathcal{G}^{\mathbb{T}}$ be such that $gx = gy$. Then $g = tf$ for some $t \in \mathcal{G}$. Furthermore,

$$c \cdot Tt \cdot Tf = c \cdot Tg = ga = tfa = tb \cdot Tf ;$$

since T preserves regular epimorphisms it follows that $c \cdot Tt = tb$, so that $t \in \mathcal{G}^{\mathbb{T}}$. Therefore $f \in \mathrm{Coequ}_{\mathcal{G}^{\mathbb{T}}}(x,y)$ and is a regular epimorphism in $\mathcal{G}^{\mathbb{T}}$.

Let now $f : (A,a) \longrightarrow (B,b) \in \mathcal{G}^{\mathbb{T}}$ be arbitrary. Let (m,p) be a regular decomposition of f in $\mathcal{G}$ and C be the domain of m. We obtain a commutative diagram:

$$\begin{array}{ccccc} TA & \xrightarrow{Tp} & TC & \xrightarrow{Tm} & TB \\ {\scriptstyle a}\downarrow & & & & \downarrow{\scriptstyle b} \\ A & \xrightarrow{p} & C & \xrightarrow{m} & B \end{array}$$

There Tp is regular, hence strong, so that $m(pa) = (b.Tm)Tp$ implies $pa = c.Tp$ for some unique $c : TC \longrightarrow C$; then also $mc = b.Tm$. Furthermore,

$$c.\eta C.p = c.Tp.\eta A = pa.\eta A = p \ ,$$

$$c.\mu C.TTp = c.Tp.\mu A = pa.\mu A = pa.Ta = c.Tp.Ta = c.Tc.TTp \ ;$$

since p, TTp are epimorphisms it follows that $c.\eta C = 1$, $c.\mu C = c.Tc$. Hence $(C,c) \in \mathcal{C}^{\mathbb{T}}$. Our diagram then shows that $m,p \in \mathcal{C}^{\mathbb{T}}$ and by the first part of the proof we have found a regular decomposition of f in $\mathcal{C}^{\mathbb{T}}$. Then the proof is completed as for 2.1.

That varieties are regular follows immediately from this theorem since the category of sets is regular and any triple thereon preserves regular epimorphisms since they are retractions.

One more transfer theorem (to sheaves) will be found in this volume. Of course it follows from 2.1 that presheaves in a regular category over any topological space or Grothendieck topology, form a regular category.

3. SUBOBJECTS; DIRECT AND INVERSE IMAGES.

Let $\mathcal{C}$ be a regular category.

1. Recall that a subobject of $A \in \mathcal{C}$ is a class of equivalent monomorphisms of codomain A. The subobject containing a monomorphism m is denoted by $\operatorname{Im} m$. A [partial] order relation between subobjects of A is defined by: $\operatorname{Im} m \leq \operatorname{Im} n$ if and only if $m = nt$ for some

$t \in G$. The intersection of a family $(\underline{x}_i)_{i\in I}$ of subobjects of A is defined as usual and denoted by $\underline{x} = \bigwedge_{i\in I} \underline{x}_i$; it is a g.l.b., i.e. $\underline{y} \leq \underline{x}$ if and only if $\underline{y} \leq \underline{x}_i$ for all i . Note that G has finite intersections. On the other hand, we define the union of $(\underline{x}_i)_{i\in I}$ as a l.u.b. (when such exists), i.e. $\underline{x} = \bigvee_{i\in I} \underline{x}_i$ in case $\underline{y} \geq \underline{x}$ if and only if $\underline{y} \geq \underline{x}_i$ for all i . (This differs from Mitchell's definition [31], but we shall soon see (3.3, below) that in the regular category G the two definitions are equivalent.) There is a greatest subobject of A , namely $1 = \mathrm{Im}\ 1_A$.

Each morphism $f \in G$ yields a subobject of its codomain: indeed all the monomorphisms m in the regular decompositions (m,p) of f form a subobject. We denote it by Im f ; if f is a monomorphism, this is indeed the subobject containing f; in general, it is an image in the sense of [31], although in this case again regularity enables us to give a definition which works as well but is somewhat more natural.

2. Each morphism $f : A \longrightarrow B \in G$ induces an '<u>inverse image</u>' map f^s defined as usual: if Im n is a subobject of B , then f^s Im n is well-defined by $f^s \mathrm{Im}\ n = \mathrm{Im}\ m$, where $fm = ng$ is a pullback. The general properties of inverse images can be found in [31] (or are easy to prove directly): $(1_A)^s$ is the identity, $(fg)^s = g^s f^s$; $f^s 1 = 1$; f^s is order-preserving, in fact preserves all existing intersections. (These properties would hold in any category with pullbacks.)

3. The existence of regular decompositions allows us to define direct images as well. If f is as above and Im m is a subobject of A , then f_s Im m is well-defined by: $f_s \mathrm{Im}\ m = \mathrm{Im}\ fm$. Equivalently, one may take a regular decomposition (n,q) of fm , and then $f_s \mathrm{Im}\ m = \mathrm{Im}\ n$.

If G is a variety, we know that every monomorphism is equivalent to precisely one inclusion map, so that the subobjects of $A \in G$ may be identified with the subalgebras of A ; when this is done it is easy to see that direct and inverse images of subobjects have their usual meaning. The same is true in an abelian category; then the direct and inverse image maps can be used for a form of diagram chasing with subobjects (the idea is due to Mac Lane [29]). This still works, to some extent, in regular categories [13] and we shall presently indicate all relevant properties.

First we show that our direct images are indeed satisfactory.

Proposition 3.1. $(1_A)_s = I$ and $(fg)_s = f_s g_s$.

Proof. The first assertion is trivial (I denotes the identity map). For the second, let Im m be a subobject such that $(fg)_s \operatorname{Im} m$ is defined. Take regular decompositions (n,q) of gm , (k,r) of fn, so that $g_s \operatorname{Im} m = \operatorname{Im} n$ and $f_s g_s \operatorname{Im} m = \operatorname{Im} k$. Then rq is a regular epimorphism (1.10) ; since $fgm = fnq = krq$, (k,rq) is a regular decomposition of fgm , whence $(fg)_s \operatorname{Im} m = \operatorname{Im} k = f_s g_s \operatorname{Im} m$.

Proposition 3.2. $f_s 1 = \operatorname{Im} f$; more generally, $f_s \operatorname{Im} g = \operatorname{Im} fg$.

Proof. This follows at once from 3.1.

Proposition 3.3. f_s is order-preserving, in fact preserves all existing unions.

Proof. Let m,n be monomorphisms with $\operatorname{Im} m \leq \operatorname{Im} n$, i.e. $m = nt$ for some t . Take regular decompositions (k,q) of fn , (ℓ,r) of qt . Then $fm = fnt = kqt = k\ell r$, so that $f_s \operatorname{Im} m = \operatorname{Im} k\ell \leq \operatorname{Im} k = f_s \operatorname{Im} n$. Hence f_s is order-preserving.

Now assume that $\underline{x} = \bigvee_{i \in I} \underline{x}_i$. By the above, $\underline{y} \geq f_s \underline{x}$ implies $\underline{y} \geq f_s \underline{x}_i$ for all i . Conversely, assume that $\underline{y} \geq f_s \underline{x}_i$ for all i . Put $\underline{y} = \operatorname{Im} n$, $\underline{x}_i = \operatorname{Im} m_i$; let (n_i, q_i) be a regular decomposition of fm_i and $fm = ng$ be a pullback. Then $\operatorname{Im} n \geq \operatorname{Im} n_i$, so that

$n_i = nt_i$ for some t_i. Then $fm_i = n_i q_i = nt_i q_i$, which in the pullback $fm = ng$ implies $m_i = mu_i$ for some u_i, i.e. $\underline{x}_i \leq \text{Im } m$. This holds for every i and therefore $\underline{x} \leq \text{Im } m$. It follows that $f_s \underline{x} \leq f_s \text{Im } m = \text{Im } fm = \text{Im } ng = n_s \text{Im } g \leq n_s 1 = \text{Im } n = \underline{y}$. Thus we have proved that $f_s \underline{x} = \bigvee_{i \in I} f_s \underline{x}_i$.

<u>Corollary</u> 3.4. $\text{Im } fg \leq \text{Im } f$, with equality if g is a regular epimorphism.

<u>Proof</u>. $\text{Im } fg = f_s \text{Im } g \leq f_s 1 = \text{Im } f$. If g is a regular epimorphism, then $\text{Im } g = 1$ and the equality holds.

4. We now investigate the relationships between direct and inverse images.

<u>Proposition</u> 3.5. Let $f : A \longrightarrow B$. For each subobject $\underline{y}$ of B, $f^s \underline{y}$ is the greatest subobject $\underline{x}$ of A such that $f_s \underline{x} \leq \underline{y}$. (In particular, $f^s \text{Im } f = 1$.)

<u>Proof</u>. Put $\underline{y} = \text{Im } n$ and let $fm = ng$ be a pullback (so that $f^s \underline{y} = \text{Im } m$). First, $f_s f^s \underline{y} = \text{Im } fm = \text{Im } ng \leq \text{Im } n = \underline{y}$. Next, let $\underline{x} = \text{Im } k$ be such that $f_s \underline{x} \leq \underline{y}$. Let (ℓ, p) be a regular decomposition of fk (so that $f_s \underline{x} = \text{Im } \ell$). Then $\text{Im } \ell \leq \text{Im } n$ and $\ell = nt$ for some t. This implies $fk = \ell p = ntp$ and, since $fm = ng$ is a pullback, $k = mu$ for some u. Hence $\underline{x} = \text{Im } k \leq \text{Im } m = f^s \underline{y}$.

<u>Corollary</u> 3.6. $f_s f^s \leq I$; $f^s f_s \geq I$; $f_s f^s f_s = f_s$; $f^s f_s f^s = f^s$.

<u>Proof</u>. The first two parts are immediate from 3.5. Next, $f_s f^s \leq I$ implies $(f_s f^s) f_s \leq f_s$, while $f^s f_s \geq I$ implies $f_s (f^s f_s) \geq f_s$; it follows that $f_s f^s f_s = f_s$. The last formula is proved similarly.

It follows from 3.6 that $f^s f_s$ is a closure operator on subobjects of the domain of f; in a variety, $f^s f_s \underline{x}$ is the subalgebra of all elements equivalent to elements of the subalgebra $\underline{x}$ modulo the congruence ker f induced by f ; in an abelian category, one finds $f^s f_s \underline{x} = \underline{x} \vee \text{Ker } f$. The other operator $f_s f^s$ is also a closure operator, but in the opposite order, and is given by:

Proposition 3.7. $f_s f^s \underline{x} = \underline{x} \wedge \text{Im } f$.

Proof. Put $\underline{x} = \text{Im } m$ and let (n,q) be a regular decomposition of f and $mn' = nm'$, $qm'' = m'q'$ be pullbacks. Then $\underline{x} \wedge \text{Im } f =$ $= \text{Im } mn'$. On the other hand, $m(n'q') = (nq)m''$ [$= fm''$] is a pullback, so that $f^s \underline{x} = \text{Im } m''$. Now the pullback axiom implies that q' is a regular epimorphism, so that (mn', q') is a regular decomposition of fm'' and $f_s f^s \underline{x} = \text{Im } mn' = \underline{x} \wedge \text{Im } f$.

Proposition 3.8. If f is a regular epimorphism then $f_s f^s = I$ (and conversely).

Proof. If f is a regular epimorphism, then $\text{Im } f = 1$ and the direct part follows at once from 3.7. If conversely $f_s f^s = I$, then $\text{Im } f = f_s 1 = f_s f^s 1 = 1$.

Proposition 3.9. If f is a monomorphism then $f^s f_s = I$.

Proof. Let $\underline{x} = \text{Im } m$ be such that $f_s \underline{x} = \text{Im } fm$ is defined. It is easily seen that $fm = (fm)1$ is a pullback (since f is a monomorphism); hence $f^s f_s \underline{x} = \text{Im } m = \underline{x}$. (This time, the converse does not generally hold.)

5. Except for 3.7, 3.8 we have not used the pullback axiom. In addition, strong decompositions could have been used instead of regular ones. This will no longer be the cases in the following sections, in which the pullback axiom is used through the (equivalent) pullback lemma which follows:

Lemma 3.10. Let $fh = gk$ be a pullback. Then $f^s \text{Im } g = \text{Im } h$ and $\text{Im } f \wedge \text{Im } g = \text{Im } fh$ (= Im gk).

Proof. We know that this is trivial when f and g are monomorphisms. In general, take regular decompositions (m,p) of f, (n,q) of g, and construct the following diagram, in which each square is a pullback:

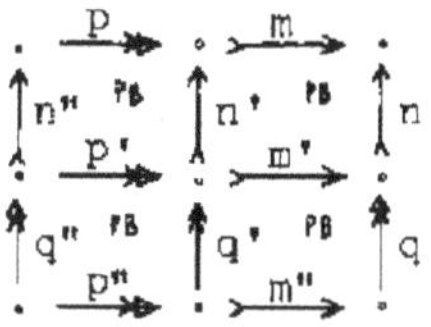

Monomorphisms and regular epimorphisms are in the diagram as indicated, due to the pullback axiom. Now juxtaposition yields a pullback $(mp)(n"q") = (nq)(m"p")$ which we may assume is $fh = gk$. Then $f^s \mathrm{Im}\, g = f^s \mathrm{Im}\, n = \mathrm{Im}\, n" = \mathrm{Im}\, h$; also, $q'p"$ is a regular epimorphism, by 1.10, so that $\mathrm{Im}\, f \wedge \mathrm{Im}\, g = \mathrm{Im}\, m \wedge \mathrm{Im}\, n = \mathrm{Im}\, mn' = \mathrm{Im}\, fh$.

Additional properties of direct and inverse images (with respect to a definition of exact sequences which can be given in any regular category, and works to a certain extent) can be found in [13]; they constitute the extension, properly said, of Mac Lane's diagram chasing with subobjects to regular categories.

4. RELATIONS.

Let $\mathcal{C}$ be a regular category.

1. If $A, B \in \mathcal{C}$, a relation $\alpha : A \longrightarrow B$ is a subobject of $A \sqcap B$. In the abelian case these are known as additive relations and have been considered most notably in [33] and, using an axiomatic approach, in [28]; if $\mathcal{C}$ is a variety, a relation is a binary relation which admits the operations.

In general, every morphism $f : A \longrightarrow B$ yields a monomorphism

$1_A \times f : A \longrightarrow A \pi B$ and a relation $\mathrm{Im}(1_A \times f)$ which can be identified with f since $\mathrm{Im}(1_A \times f) = \mathrm{Im}(1_A \times g)$ implies $f = g$ (more generally, $\mathrm{Im}(1_A \times f) \leq \mathrm{Im}(1_A \times g)$ implies $1_A \times f = (1_A \times g)t$ for some t, $f = gt$ and $1_A = 1_A t$, and $f = g$).

To each relation $\alpha : A \longrightarrow B$ corresponds a subobject $\mathrm{Im}\,\alpha = p_s \alpha$ of B (where $p : A \pi B \longrightarrow B$ is the projection); since $p_s\, \mathrm{Im}(1_A \times f) = \mathrm{Im}\, f$, the notation creates no confusion when α is a morphism. One may also write $\alpha = \mathrm{Im}(a \times b)$ (where a, b have codomains A, B respectively) and then $\mathrm{Im}\,\alpha = \mathrm{Im}\, b$.

For each relation $\alpha : A \longrightarrow B$ one has an <u>inverse</u> relation $\alpha^{-1} : B \longrightarrow A$, defined by: $\alpha^{-1} = t_s \alpha$, where $t : A \pi B \longrightarrow B \pi A$ "exchanges the components". Since $tt = 1$, $(\alpha^{-1})^{-1} = \alpha$. If $f : A \longrightarrow B$ is an isomorphism, then its inverse as a relation is $\mathrm{Im}(f \times 1_A) = \mathrm{Im}((f \times 1_A)f^{-1}) = \mathrm{Im}(1_B \times f^{-1}) = f^{-1}$ the inverse isomorphism. In general, if $\alpha = \mathrm{Im}(a \times b)$ then $\alpha^{-1} = \mathrm{Im}(b \times a)$.

We also have an order relation between relations $A \longrightarrow B$. It is clear (from 3.3) that the taking of images and inverses are order-preserving operations, and in fact preserve unions. In addition, taking inverses also preserves intersections of relations (since in the above t is an isomorphism).

2. Of course the most important operation on relations is composition, and as we discuss it we shall also give some justification for our definition of relations. There are indeed two ways of defining relations, the above and the definition in which a relation is simply a pair of morphisms $A \xleftarrow{a} R \xrightarrow{b} B$. The latter (used by [31],[25] and many others, with interesting viewpoints in [5]) allows to define an associative composition by pullbacks: i.e. if α is as above and $\beta : B \xleftarrow{b'} \cdot \xrightarrow{c} C$, then $\beta\alpha$ is given on the diagram next page, where the square is a pullback:

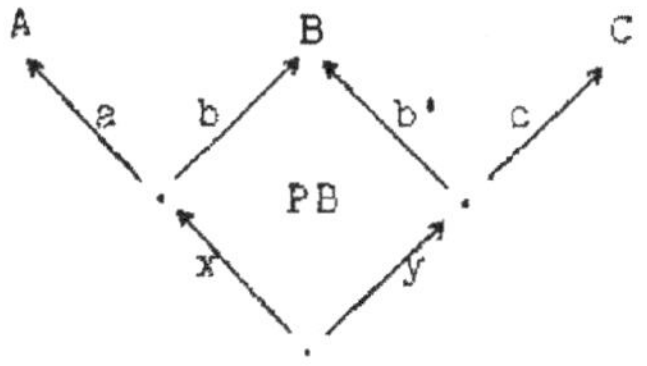

by: $\beta\alpha : A \xleftarrow{ax} \cdot \xrightarrow{cy} C$. On this definition, all properties and examples of compositions are easily available, that is, as long as one does not need, say, unions or intersections of relations; this is perfectly satisfactory, as long as extended manipulation of relations (of the kind to be found in the next part) is not needed. For example, an easy way to define intersections, or unions, of relations, using that definition, has yet to be found.

The main advantage of defining relations as subobjects is the precision one gains. In addition, unions and intersections come naturally, as well as everything in the previous section (at least, in a regular category). However, one needs a definition of composition. Composing by pullbacks has the inconvenient that when $a \times b$ and $b' \times c$ are monomorphisms in the diagram above, $ax \times cy$ need not be a monomorphism; hence if we wish to compose more than two relations, we must not assume that $a \times b$, $b' \times c$ are monomorphisms, and have to prove that composition by pullbacks yields a well-defined operation. (At the end of section 5 we show that this is not true unless the pullback axiom holds.) Another way of composing relations $\alpha : A \longrightarrow B$ and $\beta : B \longrightarrow C$ is to use Puppe's formula [33]: $\beta\alpha = r_s(p^s\alpha \wedge q^s\beta)$, where p, q, r are the projections from $A \pi B \pi C$ to $A \pi B$, $B \pi C$, $A \pi C$ respectively. Unfortunately, it is far more cumbersome to manipulate than the definition by pullbacks.

Fortunately, in a regular category, the two definitions agree. In particular, composition by pullbacks is well-defined. Incidentally, this is the first significant consequence of the pullback axiom. We state it as:

Lemma 4.1. Let $\alpha = \text{Im}(a \times b) : A \longrightarrow B$, $\beta = \text{Im}(b' \times c) : B \longrightarrow C$ be relations in the [regular] category $\mathcal{G}$, where $a \times b$ and $b' \times c$ need not be monomorphisms, and $bx = b'y$ be a pullback. Then $\text{Im}(ax \times cy) = r_s(p^s\alpha \wedge q^s\beta)$.

Proof. Let X,Y be the domains of $a \times b$, $b' \times c$. Consider the diagram

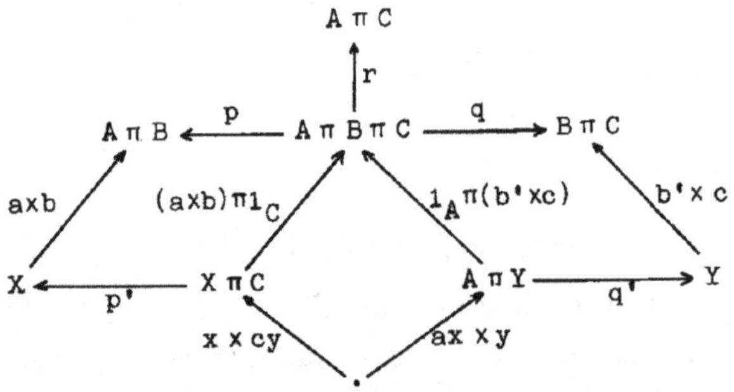

where p,q,r,p',q' are the projections. We see that the diagram commutes. In fact, the left and right squares are pullbacks, by 1.13. The same is true of the middle square. Indeed, let $x' \times c'$, $a' \times y'$ be such that $((a \times b)\pi 1_C)(x' \times c') = (1_A \pi(b' \times c))(a' \times y')$. Projecting to A,B,C yields $ax' = a'$, $bx' = b'y'$, $c' = cy'$; since $bx = b'y$ is a pullback, we have $x' = xu$, $y' = yu$ for some u. Then $x' \times c' = xu \times cyu = (x \times cy)u$, $a' \times y' = axu \times yu = (ax \times y)u$. Furthermore, $(x \times cy)u = (x \times cy)v$, $(ax \times y)u = (ax \times y)v$ implies $xu = xv$, $yu = yv$ and $u = v$ since $bx = b'y$ is a pullback.

Thus our three squares are pullbacks. Then 3.10 (a consequence of the pullback axiom) yields $p^s\alpha = p^s\text{Im}(a \times b) = \text{Im}((a \times b)\pi 1_C)$, $q^s\beta = \text{Im}(1_A \ \pi(b' \times c))$ and $p^s\alpha \wedge q^s\beta = \text{Im}(((a \times b)\pi 1_C)(x \times cy)) = \text{Im}(ax \times bx \times cy)$. Therefore $r_s(p^s\alpha \wedge q^s\beta) = \text{Im}(r(ax \times bx \times cy)) =$

$= \mathrm{Im}(ax \times cy)$, q.e.d.

The relation $\mathrm{Im}(ax \times cy) = r_s(p^s\alpha \wedge q^s\beta)$ obtained in 4.1 is now defined to be the composition $\beta\alpha$ of α and β. It is easy to see that in the abelian case (in the case of a variety) it agrees with the usual composition of additive (binary) relations.

3. We now study the properties of that operation.

Proposition 4.2. The composition of relations agrees with that of morphisms.

Proof. In the diagram

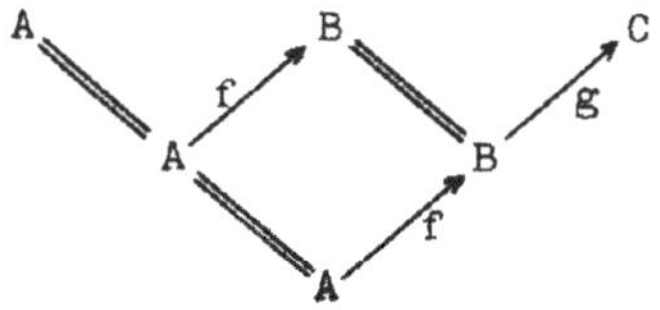

the square is a pull-back.

Proposition 4.3. The composition of relations is order-preserving.

Proof. This means that $\alpha \leq \alpha'$ and $\beta \leq \beta'$ implies $\beta\alpha \leq \beta'\alpha'$ and is clear on Puppe's formula since direct images, inverse images and intersections are order-preserving.

Proposition 4.4. The composition of relations is associative.

Proof. Consider the diagram:

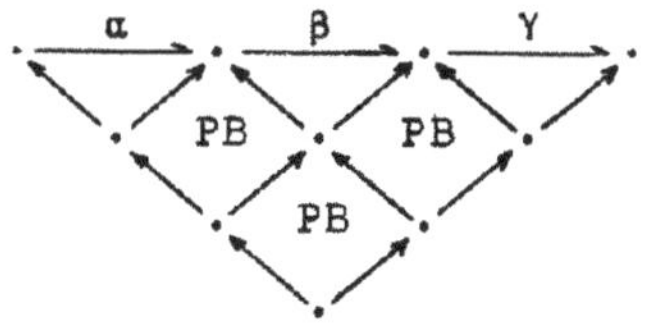

Juxtaposing the pullbacks yields pullbacks, and it follows that $\gamma(\beta\alpha)$

and $(\gamma\beta)\alpha$ are given by the same morphisms.

Identity elements are obtained by considering for each object A the diagonal $\Delta_A = 1_A \times 1_A$. Then $\varepsilon_A = \mathrm{Im}\,\Delta_A$ serves (4.5 below). (Note that $\varepsilon_A = \mathrm{Im}(1_A \times 1_A)$ can be identified with the morphism 1_A .) The notation ε means ε_A , where A is unspecified, unnamed or obvious.

Proposition 4.5. $\varepsilon\alpha = \alpha$, $\beta\varepsilon = \beta$, whenever the compositions are defined.

Proof. In the diagrams

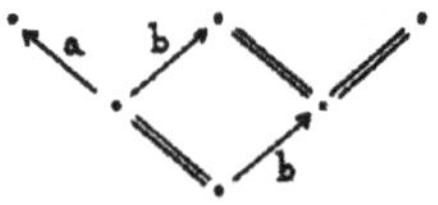

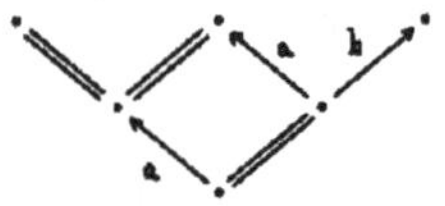

the squares are pullbacks.

When $\mathfrak{C}$ is well-powered, 4.4 and 4.5 give us a new category, whose objects are those of $\mathfrak{C}$ and morphisms are relations (in the abelian case, see [33]; in the non-abelian case, see [1]).

Proposition 4.6. $(\beta\alpha)^{-1} = \alpha^{-1}\beta^{-1}$.

Proof. Compose by pullbacks and then watch the diagram in a mirror.

Our last result shows that any relation can be obtained from two morphisms by composition:

Proposition 4.7. If $\alpha = \mathrm{Im}(a \times b)$, then $\alpha = b a^{-1}$.

Proof. [a^{-1} is the inverse of a as a relation.] In the diagram

the square is a pullback.

4. Like morphisms, relations induce functions of subobjects. To see how this is defined, we start with a lemma:

Lemma 4.8. Let $\alpha = \operatorname{Im}(a \times b)$, $\beta = \operatorname{Im}(b' \times c)$ be such that $\beta\alpha$ is defined. Then $\operatorname{Im} \beta\alpha = c_s b'^s \operatorname{Im} \alpha$.

Proof. Let $bx = b'y$ be a pullback, so that $\beta\alpha = \operatorname{Im}(ax \times cy)$. By 3.10, $\operatorname{Im} y = b'^s \operatorname{Im} b = b'^s \operatorname{Im} \alpha$; hence $\operatorname{Im} \beta\alpha = \operatorname{Im} cy = c_s b'^s \operatorname{Im} \alpha$.

The lemma shows that, for a fixed α , $\operatorname{Im} \alpha\gamma$ depends solely upon $\operatorname{Im} \gamma$. In particular, if $\alpha : A \longrightarrow B$ and $\operatorname{Im} m$ is a subobject of A, then $\operatorname{Im} \alpha m$ does not depend on the choice of m in the subobject and hence may be denoted by $\alpha_s \operatorname{Im} m$, thereby defining α_s . The following properties are then immediate from the lemma:

Proposition 4.9. If $\alpha = \operatorname{Im}(a \times b)$ then $\alpha_s = b_s a^s$; $\operatorname{Im} \alpha\gamma = \alpha_s \operatorname{Im} \gamma$; if α is a morphism then α_s has the same meaning as before.

Further properties of this new kind of direct image are given by:

Proposition 4.10. a) α_s is order-preserving; b) $(\alpha\beta)_s = \alpha_s\beta_s$; c) $\epsilon_s = I$; d) $\alpha_s 1 = \operatorname{Im} \alpha$; e) $\alpha \leq \beta$ implies $\alpha_s \leq \beta_s$.

Proof. a) and c) are clear from 4.9. Also, $(\alpha\beta)_s \operatorname{Im} m = \operatorname{Im} \alpha\beta m = \alpha_s \operatorname{Im} \beta m = \alpha_s\beta_s \operatorname{Im} m$, which proves b) . If $\alpha = \operatorname{Im}(a \times b)$ then $\alpha_s 1 = b_s a^s 1 = b_s 1 = \operatorname{Im} b = \operatorname{Im} \alpha$, which proves d) . Finally, assume that $\alpha \leq \beta$; write $\alpha = \operatorname{Im}(a \times b)$, $\beta = \operatorname{Im}(a' \times b')$, where $a \times b$ and $a' \times b'$ are monomorphisms, so that $a \times b = (a' \times b')t$ for some t . By 3.6, $t_s t^s \leq I$; hence $\alpha_s = b_s a^s = b'_s t_s t^s a'^s \leq b'_s a'^s = \beta_s$.

We may also define inverse images by: $\alpha^s = (\alpha^{-1})_s$. The following properties are then immediate from 4.9, 4.10:

Proposition 4.11. a) if α is a morphism, then α^s has the same meaning as before; b) if $\alpha = \mathrm{Im}(a \times b)$ then $\alpha^s = a_s b^s$; c) $(\alpha^{-1})^s = \alpha_s$, $(\alpha^{-1})_s = \alpha^s$; d) α^s is order-preserving; e) $(\alpha\beta)^s = \beta^s\alpha^s$; f) $\epsilon^s = I$; g) $\alpha \leq \beta$ implies $\alpha^s \leq \beta^s$.

In the case when $\mathfrak{G}$ is a variety, let $A, B \in \mathfrak{G}$ and R be a subalgebra of $A \sqcap B$. If $a : R \longrightarrow A$, $b : R \longrightarrow B$ are defined by: $a : (x,y) \longmapsto x$, $b : (x,y) \longmapsto y$, then the relation α which corresponds to R is $\mathrm{Im}(a \times b)$. If S is a subalgebra (= subobject) of A, then we can interpret $\alpha_s S$ as follows: first $a^s S = \{(x,y) \in R ; x \in S\}$ and then $\alpha_s S = b_s a^s S = \{ y \in B ; (x,y) \in R$ for some $x \in S \}$. Thus $\alpha_s S$ has the usual meaning. The same is true for inverse images.

Note that 4.9, 4.10, 4.11 extend to relations the properties of direct and inverse images under morphisms, except when it comes to preserving unions or intersections. The case of a variety shows that one cannot expect properties of that kind since e.g. direct images under a relation do not preserve unions of subalgebras (unless they happen to be set-theoretical unions).

5. We now give criteria to recognize morphisms among relations.

Proposition 4.12. Let $\alpha : A \longrightarrow B$ be a relation. The following are equivalent:

i) α is a morphism;
ii) $\alpha\alpha^{-1} \leq \epsilon_B$ and $\alpha^{-1}\alpha \geq \epsilon_A$;
iii) $\alpha\alpha^{-1} \leq \epsilon_B$ and $\alpha^s 1 = 1$.

Proof. If first $\alpha = f$ is a morphism, then $f^{-1} = \mathrm{Im}(f \times 1_A)$ and since $1_A 1_A = 1_A 1_A$ we conclude that $f f^{-1} = \mathrm{Im}(f \times f) = \mathrm{Im}\, \Delta_B f$. Hence $f f^{-1} \leq \mathrm{Im}\, \Delta_B = \epsilon_B$. On the other hand, let $fx = fy$ be a pullback, so that $f^{-1} f = \mathrm{Im}(x \times y)$. Since $f 1_A = f 1_A$, we have $1_A = xt = yt$ for some t, and therefore $\Delta_A = (x \times y)t$; hence $f^{-1} f \geq \mathrm{Im}\, \Delta_A = \epsilon_A$. Thus

i) implies ii) . It is clear that ii) implies iii) .

Conversely, assume that iii) holds. Put $\alpha = \mathrm{Im}(a \times b)$, where $a \times b$ is a monomorphism. Then $\mathrm{Im}\, a = \mathrm{Im}\, \alpha^{-1} = \alpha^{s}1 = 1$, so that a is a regular epimorphism. On the other hand, let $ax = ay$ be a pullback. Then $\mathrm{Im}(bx \times by) = \alpha\alpha^{-1} \leq \mathrm{Im}\, \Delta_B$, whence $bx \times by = \Delta_B t$ for some t , i.e. $bx = t = by$. Since $ax = ay$, it follows that $(a \times b)x = (a \times b)y$ and $x = y$. But $a \in \mathrm{Coequ}(x,y)$, since it is a regular epimorphism, and therefore a is an isomorphism. Hence $\alpha = b a^{-1}$ is a morphism.

Proposition 4.13. Let f be a morphism. Then f is a monomorphism if and only if $f^{-1}f = \epsilon$, and a regular epimorphism if and only if $f f^{-1} = \epsilon$.

Proof. First, f is a monomorphism if and only if $f1 = f1$ is a pullback, if and only if $f^{-1}f = \mathrm{Im}\, \Delta$. If f is a regular epimorphism, then $f f^{-1} = \mathrm{Im}(f \times f) = \mathrm{Im}\, \Delta f = \mathrm{Im}\, \Delta$; if conversely $f f^{-1} = \epsilon$, then $\mathrm{Im}\, f = f_s 1 = f_s f^s 1 = \epsilon_s 1 = 1$ and f is a regular epimorphism.

Note that the first half of 3.6, as well as 3.8, 3.9 follow from these results (in fact, 4.13 is more accurate that 3.8-3.9).

The importance of 4.13 is that in a regular category it provides the only criterion for recognizing monomorphisms that can be manipulated in much the same way kernels are manipulated in abelian categories. This has a great deal to do with the nature of the proofs in the next part.

In an abelian category, much better criterions exist to recognize morphisms: for instance, $\alpha^{s}1 = 1$, $\alpha_s 0 = 0$ [33],[29]. This criterion is still valid in the variety of all groups, and in that of rings, but not in general (even if there is a zero object; monoids, even sets with a base point, provide easy counterexamples). In certain cases (semi-groups, with perhaps an identity and/or a zero), ii) may be weakened into: $\alpha_s\alpha^s \leq I$, $\alpha^s\alpha_s \geq I$ [an unpublished result of the author, which brings some improvements to [13]].

6. Another type of direct or inverse image is finally obtained from any morphism $f : A \longrightarrow B$, using relations $\alpha : A \longrightarrow A$ or $\beta : B \longrightarrow B$. Namely, we define $f_r\alpha = f\alpha f^{-1} : B \longrightarrow B$, and $f^r\beta = f^{-1}\beta f : A \longrightarrow A$.

A direct study of these maps is not necessary since they can be reduced to ordinary direct or inverse images. Define a functor $\tilde{} : G \longrightarrow G$ by: $\tilde{A} = A \pi A$, $\tilde{f} = f \pi f$. This functor is nice; namely:

Proposition 4.14. The tilda functor is left exact and preserves regular decompositions.

Proof. It already preserves monomorphisms, regular epimorphisms (by 1.12) and finite products. Finally, let $m \in Equ(f,g)$. Then $\tilde{f}\tilde{m} = \tilde{g}\tilde{m}$ Also, $\tilde{f}(h \times k) = \tilde{g}(h \times k)$ implies: $fh = gh$, $fk = gk$; $h = mt$, $k = mu$ for some t,u; and $h \times k = \tilde{m}(t \times u)$. Since $\tilde{m}$ is a monomorphism it follows that $\tilde{m} \in Equ(\tilde{f},\tilde{g})$. Thus $\tilde{}$ preserves equalizers; this completes the proof.

Using the tilda functor, we can reduce f_r, f^r to previously studied direct or inverse images; namely:

Proposition 4.15. $f_r = \tilde{f}_s$, $f^r = \tilde{f}^s$.

Proof. Take $\alpha = Im(a \times b) : A \longrightarrow A$. We also have $f = Im(1 \times f)$, $f^{-1} = Im(f \times 1)$ and since $1a = a1$, $b1 = 1b$ and $11 = 11$ are pullbacks it follows that $f\alpha f^{-1} = Im(fa \times fb) = Im\,\tilde{f}(a \times b) = \tilde{f}_s\alpha$.

Let now $\beta = Im(c \times d) : B \longrightarrow B$. Let $fc' = cf'$, $fd' = dg'$ and $f'g'' = g'f''$ be pullbacks, so that $f^{-1}\beta f = Im(c'g'' \times d'f'')$. We now consider the following commutative diagram:

$$\begin{array}{ccccc} X & \xrightarrow{\Delta} & \bar{X} & \xrightarrow{c \sqcap d} & \tilde{B} \\ {\scriptstyle f'g''} \uparrow & & \uparrow {\scriptstyle f' \sqcap g'} & & \uparrow {\scriptstyle \tilde{f}} \\ T & \xrightarrow[g'' \times f'']{} & Y \sqcap Z & \xrightarrow[c' \sqcap d']{} & \tilde{A} \end{array}$$

where X, Y, Z, T are the respective domains of $c \times d$, c', d', $g'' \times f''$. The left square is a pullback. Indeed $(f' \sqcap g')(u \times v) = \Delta w$ implies $f'u = w = g'v$, $u = g''t$ and $v = f''t$ for some t and $u \times v = (g'' \times f'')t$, $w = f'g''t$; the factorization is unique since $g'' \times f''$ is a monomorphism (as $f'g'' = g'f''$ is a pullback). In addition, the right-hand square is also a pullback, since products preserve pullbacks. It follows that $\tilde{f}\,((c' \sqcap d')(g'' \times f'')) = ((c \sqcap d)\Delta)(f'g'')$ is a pullback; i.e. $\tilde{f}(c'g'' \times d'f'') = (c \times d)(f'g'')$ is a pullback. Then 3.10 yields $\tilde{f}^s \beta = \tilde{f}^s \operatorname{Im}(c \times d) = \operatorname{Im}(c'g'' \times d'f'') = f^r \beta$, q.e.d.

In particular, it follows from 4.15 that f_r (f^r) preserves existing unions (intersections), a fact which would not be easy to prove directly.

7. With all this we can account for just about all the elementary properties of relations and morphisms which hold in, say, a variety. The other model of relation theory known to us (Mac Lane's [ ]) is definitely oriented towards the abelian case and therefore the above does not fit into it. For one thing, even when $\mathfrak{a}$ is well-powered and the subobjects of a given $A \in \mathfrak{a}$ do form a lattice (i.e. when $\mathfrak{a}$ has finite unions), then this lattice need not be modular. For another, it is not true that $\alpha \alpha^{-1} \alpha = \alpha$ always holds, although this is the case in an abelian category, and also in the varieties of groups and of rings, and, in general, when either α or α^{-1} is a morphism; counterexamples can easily be found with sets. If allowances are made for this, a number of results in [28] will still hold in our situation (though not the finer ones). Except for the characterization of morphisms, the more elementary aspects are saved.

5. CONGRUENCES.

1. Throughout, $\mathcal{C}$ denotes a given regular category.

If $f : A \longrightarrow B$ is a morphism, the relation $f^{-1}f : A \longrightarrow A$ is the congruence induced by f ; it will be denoted by $\ker f$. One can calculate $f^{-1}f$ by pullbacks: if $fx = fy$ is a pullback then $\ker f = f^{-1}f = \mathrm{Im}(x \times y)$; note that $x \times y$ is a monomorphism since $fx = fy$ is a pullback. Of course (x,y) is the kernel pair of f . A congruence is any relation of the form $\ker f$. We note that, if (m,p) is a regular decomposition of f , then f and p have same kernel pair, so that $\ker f = \ker p$; hence any congruence is induced by some regular epimorphism.

For instance, ϵ_A is a congruence on A (since $\epsilon_A = \ker 1_A$); it follows from 4.12 that it is in fact the least congruence on A. There also exists a greatest congruence on A. Indeed $\mathcal{C}$ is finitely complete, so that there exists a null object N of $\mathcal{C}$ (such that for every $X \in \mathcal{C}$ there exists precisely one morphism $n_X : X \longrightarrow N$) : namely, the limit of the empty diagram. If $f : A \longrightarrow B$ is any morphism of domain A , then $n_B f = n_A$, so that $n_A^{-1}n_A = f^{-1}n_B^{-1}n_B f \geq f^{-1}f$; thus $\upsilon_A = \ker n_A$ is the greatest congruence on A . In fact, we see that $n_A p = n_A q$, where $p, q : A \pi A \longrightarrow A$ are the projections, is a pullback; therefore $\upsilon_A = \mathrm{Im}\, \tilde{1}_A$ is the greatest subobject of $A \pi A$.

In a variety, the definition we used for $\ker f$ yields the congruence induced by f in the usual sense. In an abelian category, there is no need to use congruences because the congruence $\ker f$ induced by f gives us no more information than the kernel $\mathrm{Ker}\, f$ of f . Precisely:

Proposition 5.1. Assume that $\mathcal{C}$ is an abelian category. Then:

a) $\mathrm{Ker}\ f = (\ker f)_s 0$; b) if α, β are congruences on $A \in \mathcal{G}$ with $\alpha_s 0 = \beta_s 0$, then $\alpha = \beta$.

Remark. In other words, each of Ker f , ker f is completely determined by the other. This expresses the well-known fact that in an abelian group a congruence is completely determined by the class of the identity element.

Proof. For a), take $k \in \mathrm{Ker}\ f$; since $fk = 00$ is a pullback we see that $\mathrm{Ker}\ f = f^s 0$. We have to relate this to $(\ker f)_s 0$, which can be written as $y_s x^s 0$, where $fx = fy$ is a pullback (so that $\ker f = \mathrm{Im}(x \times y)$). First, $f_s y_s x^s 0 = f_s x_s x^s 0 \leq f_s 0 = 0$, so that $y_s x^s 0 \leq f^s 0 = \mathrm{Ker}\ f$. Conversely, $f0 = fk$ implies $k = yt$, $xt = 0$ for some t ; then $x_s\ \mathrm{Im}\ t = 0$, whence $\mathrm{Im}\ t \leq x^s 0$ and $\mathrm{Ker}\ f = \mathrm{Im}\ k = y_s\ \mathrm{Im}\ t \leq y_s x^s 0$. (In fact, we see that a) will hold even if $\mathcal{G}$ is not abelian, as long as every object of $\mathcal{G}$ has a least subobject 0).

If now α and β are congruences on the same object with $\alpha_s 0 = \beta_s 0$, then we can write $\alpha = \ker f$, $\beta = \ker g$, where f and g are [regular] epimorphisms; by part a) , we then have Ker f = Ker g , and this implies that $f = tg$ for some isomorphism t , whence ker f = ker g . This completes the proof.

In the two basic examples of regular categories, there is another property related to congruences. We know that in a variety any relation [admitting the operations] which is reflexive, symmetric and transitive, is a congruence. In general, a congruence α in any regular category is reflexive (i.e. $\varepsilon \leq \alpha$); symmetric (i.e. $\alpha^{-1} = \alpha$), for if $fx = fy$ is a pullback, then so is $fy = fx$, so that $x \times y$ and $y \times x$ are equivalent monomorphisms and $(\ker f)^{-1} = \mathrm{Im}(y \times x) = \mathrm{Im}(x \times y) = \ker f$; and transitive (i.e. $\alpha\alpha \leq \alpha$; equivalently, since α is reflexive,

$\alpha\alpha = \alpha$), since $\alpha = \ker f = f^{-1}f$ for some regular epimorphism f and then $\alpha\alpha = f^{-1}ff^{-1}f = f^{-1}f$ by 4.13 . However, the converse just might not be true. The condition

(L) Every relation $A \longrightarrow A$ which is reflexive, symmetric and transitive is a congruence

will be called Lawvere's condition; it is equivalent to one of the conditions in Lawvere's theorem characterizing finitary varieties [25] . We have seen that it holds in any variety; in addition:

Proposition 5.2. Every abelian category satisfies Lawvere's condition.

Proof. Let α be a reflexive, symmetric and transitive relation on the object A of an abelian category $\mathcal{G}$; pick a monomorphism $k \in \alpha_s 0$ and an exact sequence $0 \longrightarrow \cdot \xrightarrow{k} \cdot \xrightarrow{f} \cdot \longrightarrow 0$; it suffices to prove that $\alpha = \ker f$. If $\mathcal{G}$ is the category of all R-modules, where R is some ring, then since $\mathcal{G}$ is a variety α is a congruence and then $\alpha = \ker f$ is true by 5.1. In the general case we observe that the conditions on α and f and the conclusion that $\alpha = \ker f$ can be expressed in terms of finitely many objects and morphisms of $\mathcal{G}$; hence Mitchell's full embedding theorem can be used to go back to the particular case of R-modules.

2. In a regular category, congruences are manipulated much as in a variety.

Proposition 5.3. $\ker fg \geq \ker g$, with equality if f is a monomorphism.

Proof. $\ker fg = g^{-1}f^{-1}fg \geq g^{-1}g = \ker g$ by 4.12 ; if f is a monomorphism, the equality follows from 4.13.

A converse of 5.3 is the following "induced homomorphism theorem":

Proposition 5.4. If $\ker f \leq \ker g$ and f is a regular epimorphism, then $g = tf$ for some t; t is a monomorphism if and only if $\ker f = \ker g$.

Proof. Let $fx = fy$, $gx' = gy'$ be pullbacks; then $\mathrm{Im}(x \times y) \leq \mathrm{Im}(x' \times y')$, which implies (since $x \times y$, $x' \times y'$ are monomorphisms), $x \times y = (x' \times y')u$ for some u. Hence $gx = gx'u = gy'u = gy$; since $f \in \mathrm{Coequ}(x,y)$, it follows that $g = tf$ for some (unique) t. If furthermore $\ker f = \ker g$, then, by 4.13:

$$t^{-1}t = f\,f^{-1}t^{-1}t\,f\,f^{-1} = f\,g^{-1}\,g\,f^{-1} = f\,f^{-1}\,f\,f^{-1} = \epsilon$$

and t is a monomorphism; the converse follows from 5.3.

A first consequence of 5.4 is that, if f and g both are regular epimorphisms, then $\ker f = \ker g$ implies $g = tf$, where t is an isomorphism by 1.11, 1.8; if conversely f and g are equivalent, then $\ker f = \ker g$ by 5.3. Hence there is a one-to-one correspondance between the regular quotient-objects of a given $A \in \mathcal{C}$ and the congruences on A.

In particular

Corollary 5.5. A well-powered regular category is also regularly co-well-powered.

The next basic operations are direct and inverse images of congruences under morphisms.

Proposition 5.6. For any morphism f and congruence α, $f^r\alpha$ (if defined) is a congruence. Namely, $f^r \ker g = \ker gf$. In particular, $f^r\epsilon = \ker f$.

Proof. $f^r \ker g = f^{-1}g^{-1}\,g\,f = \ker gf$.

Predictably, direct images do not work so well. However:

<u>Proposition</u> 5.7. Let f be a morphism and α be a congruence on the domain of f. If f is a regular epimorphism and $\alpha \geq \ker f$, then $f_r\alpha$ is a congruence.

<u>Proof</u>. Put $\alpha = \ker g$. By 5.4, $g = tf$ for some t. By 4.13, $f_r\alpha = f\,g^{-1}g\,f^{-1} = f\,f^{-1}t^{-1}t\,f\,f^{-1} = t^{-1}t = \ker t$.

<u>Proposition</u> 5.8. $f_r \ker f \leq \varepsilon$, $f_r\varepsilon \leq \varepsilon$; in each case, the equality holds if and only if f is a regular epimorphism.

<u>Proof</u>. We always have $f_r\varepsilon \leq f_r \ker f = f\,f^{-1}\,f\,f^{-1} \leq \varepsilon$ by 4.12. By 4.13, $f_r\varepsilon = f\,f^{-1} = \varepsilon$ if and only if f is a regular epimorphism, and then $f_r \ker f = \varepsilon$. If conversely $f_r \ker f = \varepsilon$ then let (m,p) be a regular decomposition of f; then $\operatorname{Im} \Delta = \varepsilon \leq f_r \ker f \leq \operatorname{Im} \tilde{f} = \operatorname{Im} \tilde{m}$, so that $\Delta = \tilde{m}(u \times v)$ for some u,v; then $mu = 1$ and m (and hence f) is a regular epimorphism. [One can also show that $f\,f^{-1}f = f$ for all f, so that $f_r \ker f = f_r\varepsilon$.]

This last proof shows that in general $f_r\alpha$ cannot be a congruence if f is not a regular epimorphism.

3. If $\mathcal{C}$ is regularly co-well-powered, then for each congruence α on $A \in \mathcal{C}$ we can select a regular epimorphism f with $\ker f = \alpha$ and thus select an object of $\mathcal{C}$ (the codomoain of f) which may be called A/α. If $\mathcal{C}$ is not regularly co-well-powered then one may still be willing to use the notation A/α despite the fact that it can only denote an object defined up to isomorphism only. At any rate we may now wonder if the isomorphisms theorems which hold in a variety can still be formulated in a regular category. The answer is yes, although the first two ($\operatorname{Im} f \cong A/\ker f$ and $(A/\alpha)/(\beta/\alpha) \cong A/\beta$) are now devoid of mathematical content (i.e. are trivial), the only difficulty being to set-up the obvious appropriate definitions. The last one, however, is still of interest.

First, let α be a congruence on $A \in \mathcal{G}$. If $m : B \longrightarrow A$ is a monomorphism, then the restriction $\alpha_{|B}$ of α to B may be defined by: $\alpha_{|B} = m^r\alpha$. The extension of B under α may be defined as the domain $\alpha(B)$ of a monomorphism n such that $\operatorname{Im} n = \alpha_s \operatorname{Im} m$.(In case $\mathcal{G}$ is a variety, this will yield the usual restriction, and the union of all classes modulo α of the elements of B, respectively). The last isomorphism theorem may then be stated as:

Proposition 5.9. Let $m : B \longrightarrow A$ be a monomorphism and α be a congruence on A. Put $\alpha' = \alpha_{|B}$, $C = \alpha(B)$, $\alpha'' = \alpha_{|C}$. Then $B/\alpha' \cong C/\alpha''$.

Proof. Let C be the domain of a monomorphism n such that $\operatorname{Im} n = \alpha_s \operatorname{Im} m$; then $\alpha'' = n^r\alpha$, $\alpha' = m^r\alpha$. First we note that $\operatorname{Im} n \geq \epsilon_s \operatorname{Im} m = \operatorname{Im} m$, so that $m = nt$ for some (monomorphism) t . Let p,q be regular epimorphisms such that $\alpha' = \ker p$, $\alpha'' = \ker q$ (so that the codomains of p,q may be denoted by B/α' , C/α''). We have $\ker qt = t^r n^r \alpha = m^r\alpha = \ker p$; hence by 5.4 $qt = up$ for some monomorphism $u : B/\alpha' \longrightarrow C/\alpha''$. We want u to be an isomorphism, and it suffices to show that it is a regular epimorphism.

First

$$\operatorname{Im} \alpha'' t = \operatorname{Im} n^{-1}\alpha n t = \operatorname{Im} n^{-1}\alpha m = n^s \operatorname{Im} \alpha m = n^s n_s 1 = 1 .$$

Hence $q^s \operatorname{Im} qt = \operatorname{Im} q^{-1}qt = \operatorname{Im} \alpha'' t = 1$. Since q is a regular epimorphism, it follows that $\operatorname{Im} qt = q_s q^s \operatorname{Im} qt = q_s 1 = 1$. i.e. qt is a regular epimorphism. By 1.9, so is u , q.e.d.

In a variety, 5.9 reduces to the usual isomorphism theorem (as stated e.g. in [ ]); e.g. in the variety of groups it means that $HK/K \cong H/H \cap K$ whenever H,K are subgroups of a group G with K normal.

The "correspondance theorem" which is sometimes included in the first isomorphism theorem, also retains some interest; in this case, it says that a regular epimorphism $f : A \longrightarrow B$ induces a one-to-one correspondance, which is order-preserving (both ways) between the congruences on B and the congruences on A that contain $\gamma = \ker f$. The maps are of course f_r, f^r (5.6, 5.7); for each congruence β on B, $f_r f^r \beta = \beta \wedge \operatorname{Im} \tilde{f} = \beta$; for each congruence $\alpha \geq \gamma$ on A, $f^r f_r \alpha = f^{-1} f \alpha f^{-1} f = \gamma\alpha\gamma$ is $\geq \alpha$ since $\gamma \geq \epsilon$, and $\leq \alpha\alpha\alpha = \alpha$ since $\gamma \leq \alpha$; thus this, too, extends to regular categories.

4. We now have completed the basic study of subobjects, relations and congruences in a regular category, and shall give some evidence that the same body of properties would not be kept if the axioms of regular categories were substantially weakened. We do not have a precise necessary and sufficient condition to that effect, but can make the following remarks.

The assumption of finite completeness cannot be weakened since we need products to describe relations and pullbacks to compose them, as well as for inverse images and intersections. (One should note that it <u>can</u> be somewhat weakened if no hard manipulation of relations is necessary, as, for instance, in Barr's contribution to this volume.)

The assumption that there exist regular decompositions is mild, in view of 1.6. Yet there is no doubt that if one needs only "nice" decompositions it would be possible (hence, preferable) to start with strong decompositions; the greater part of section 3 would still hold (even without the pullback axiom). However, if we wish to account for basic algebraic phenomena, we certainly cannot overlook the induced homomorphism theorem (first part of 5.4). This does not have to be formulated with congruences (kernel pairs will do as nicely) but no matter what formulation is chosen, the property requires that our chosen epimorphisms be at least subregular. This means that we have to start with regular decompositions (1.3). For this reason, nothing can be gained by considering factorization systems (still granted that we want a suitable categorical description of basic algebraic phenomena), since the decompositions therein will have to be regular and then no gain of generality will occur.

The pullback axiom now has the effect of ruling out a number of topological examples and is more difficult to justify at this level. To do this, we shall refer the reader to the discussion in paragraph 4.2 and show that, in a category with regular decompositions, in which the pullback axiom does not hold, the composition of relations by pullbacks is not well-defined: that is, if $bx = b'y$ is a pullback, $\mathrm{Im}(ax \times cy)$ does not depend solely upon $\mathrm{Im}(a \times b)$ and $\mathrm{Im}(b' \times c)$. To see this, let $pf' = fp'$ be a pullback, where p is a regular epimorphism and p' is not; p' has a regular decomposition (n,q) in which n is not an isomorphism. Let $\alpha = \epsilon = \mathrm{Im}(1 \times 1)$, $\beta = f^{-1} = \mathrm{Im}(f \times 1)$; note that we also have $\alpha = \mathrm{Im}(p \times p)$. Calculating $\beta\alpha$ by pullbacks, from $\alpha = \mathrm{Im}(1 \times 1)$, yields $\mathrm{Im}(f \times 1) = \beta$. If we use $\alpha = \mathrm{Im}(p \times p)$, we obtain $\mathrm{Im}(pf' \times p') = \mathrm{Im}(fp' \times p') = \mathrm{Im}((f \times 1)nq) = \mathrm{Im}((f \times 1)n)$, and this is a different relation since otherwise n would have to be an isomorphism.

Of course there are other approaches to algebraic phenomena and we certainly do not advocate the above as a panacea. Yet in some cases regular categories have a definite advantage in both generality and precision.

6. LIMITS AND COLIMITS IN A REGULAR CATEGORY.

1. In a variety limits and colimits can be constructed in terms of elements. In this section we give similar constructions in an arbitrary regular category $\mathcal{C}$ and give a number of related facts and applications.

For the notation, we know that a limit or colimit is the limit or colimit of a diagram (=functor) $\mathcal{D} : I \longrightarrow \mathcal{C}$, where I is a small category; for all objects $i \in I$ and morphisms $m \in I$, we write D_i for $\mathcal{D}(i)$ and D_m for $\mathcal{D}(m)$; morphisms of diagrams over I will be denoted by $\boldsymbol{a} = (a_i)_{i \in I}$ or by similar notations; instead of using constant diagrams (which is cumbersome) we denote the limit of $\mathcal{D}$ by $(L, (\ell_i)_{i \in I}$, and similarly for the colimits, when they exist, and use a similar notation for compatible (cocompatible) families (= morphisms from (to) constant diagrams).

2. First of all, in any (= not necessarily regular) complete category, there already is an "elementary" construction of limits. Namely, notation being as above, let $P = \prod_{i \in I} D_i$ be the product, with projections $p_i : P \longrightarrow D_i$; let $k \in \bigwedge_{m \in I} \mathrm{Equ}(p_j, D_m p_i$, where in the intersection $m : i \longrightarrow j$; put $k : L \longrightarrow P$ and $\ell_i = p_i k$. Then $(L, (\ell_i)_{i \in I})$ is a limit of $\mathcal{D}$ [9].

If $\mathcal{C}$ is regular (and complete), we have the following property:

<u>Proposition</u> 6.1. With the same notation, let $(A, (a_i)_{i \in I})$ be a compatible family for $\mathcal{D}$ inducing $a : A \longrightarrow L$. Then $\ker a = \bigwedge_{i \in I} \ker a_i$. In particular, $\bigwedge_{i \in I} \ker \ell_i = \varepsilon$.

<u>Proof</u>. We first prove the property in case I is discrete; i.e. $(a_i)_{i \in I}$ is just a family of morphisms $A \longrightarrow D_i$ and $a = \chi_{i \in I}\, a_i$.

Let $a_i x_i = a_i y_i$ be a pullback, so that $\ker a_i = \mathrm{Im}(x_i \times y_i)$; similarly, $\ker a = \mathrm{Im}(x \times y)$, where $ax = ay$ is a pullback; we have to show that $\mathrm{Im}(x \times y) = \bigwedge_{i \in I} \mathrm{Im}(x_i \times y_i)$. First, $a_i = p_i a$ [where p_i is the projection from the product] so that $\ker a \leq \ker a_i$ for all i . Conversely, assume that $\mathrm{Im}(u \times v) \leq \mathrm{Im}(x_i \times y_i)$ for all i. Since $x_i \times y_i$ is a monomorphism, we then have $u \times v = (x_i \times y_i) t_i$ for some t_i . It follows that $p_i au = a_i x_i t_i = a_i y_i t_i = p_i av$ for all i, whence $au = av$; hence $u \times v = (x \times y)t$ for some t and $\mathrm{Im}(u \times v) \leq \mathrm{Im}(x \times y)$. Thus the formula is proved in that case.

In the general case, we have (keeping the same notation) $a_i = \ell_i a = p_i ka$ for all i , so that $ka = \bigtimes_{i \in I} a_i$. Hence

$$\bigwedge_{i \in I} \ker a_i = \ker \bigtimes_{i \in I} a_i = \ker ka = \ker a \ .$$

In any regular category we also have the following description of equalizers:

<u>Lemma</u> 6.2. Let m be a monomorphism. The following are equivalent: i) $m \in \mathrm{Equ}(f,g)$; ii) $\mathrm{Im}(m \times m) = g^{-1}f \wedge \epsilon$; iii) $\mathrm{Im}\, m = \Delta^s(g^{-1}f)$. In particular $\mathrm{Equ}(f,g) = \Delta^s(g^{-1}f)$.

<u>Proof</u>. Let $fx = gy$ be a pullback; then $x \times y$ is a monomorphism and calculating $g^{-1}f$ by pullbacks yields $g^{-1}f = \mathrm{Im}(x \times y)$. If $(x \times y)k = \Delta n$ is another pullback, then $\mathrm{Im}(n \times n) = \mathrm{Im}\,\Delta n = g^{-1}f \wedge \epsilon$ and $\mathrm{Im}\, n = \Delta^s(g^{-1}f)$. We now show that $n \in \mathrm{Equ}(f,g)$. First, $xk = yk = n$, so that $fn = fxk = gyk = gn$; also, n is a monomorphism. Further assume that $fh = gh$. Then $h = xt = yt$ in the first pullback, whence $\mathrm{Im}(h \times h) \leq \mathrm{Im}(x \times y)$; since also $\mathrm{Im}(h \times h) = \mathrm{Im}\,\Delta h \leq \epsilon$, it follows that $\mathrm{Im}(h \times h) \leq \mathrm{Im}(n \times n)$; since $n \times n$ is a monomorphism, we conclude that $h \times h = (n \times n)u$, and $h = nu$ for some (unique) u . Thus $n \in \mathrm{Equ}(f,g)$.

Now, for any monomorphism m , each of i), ii), iii) is equivalent to $\mathrm{Im}\, m = \mathrm{Im}\, n$, as readily seen, so that these conditions are equivalent.

3. We now turn to colimits. In this case, we can give a new (= not due to [9]) construction. It is based upon the following lemma:

Lemma 6.3. $cf = cg$ if and only if $\mathrm{Im}(f \times g) \leq \ker c$.

Proof. Let $cx = cy$ be a pullback. Then $cf = cg$ is successively equivalent to: $f = xt$, $g = yt$ for some t ; $f \times g = (x \times y)t$ for some t ; $\mathrm{Im}(f \times g) \leq \mathrm{Im}(x \times y) = \ker c$.

Proposition 6.4. Let: $\mathcal{D} : I \longrightarrow \mathcal{C}$ be a diagram over the small category I ; $S = \coprod_{i \in I} D_i$ be a coproduct, with injections $n_i : D_i \longrightarrow S$; $c : S \longrightarrow C$ be a morphism. Then $(C, (cn_i)_{i \in I})$ is a colimit of $\mathcal{D}$ if and only if c is a regular epimorphism and $\ker c$ is the least congruence on S containing $\mathrm{Im}(n_i \times n_j D_m)$ for all $m : i \longrightarrow j \in I$.

Remark. The result has to be stated that way since we cannot be sure that there will be a least congruence containing all $\mathrm{Im}(n_i \times n_j D_m)$ (even if the coproduct exists); in fact we shall use 6.4 later to produce such least congruences.

Proof. First, assume that $(C, (cn_i)_{i \in I})$ is a colimit of $\mathcal{D}$. Then in particular it is a cocompatible family, so that $cn_i = cn_j D_m$ whenever $m : i \longrightarrow j \in I$; hence $\ker c$ contains all $\mathrm{Im}(n_i \times n_j D_m)$, by the lemma. If furthermore $\ker f$ is a congruence on S with that property, then $fn_i = fn_j D_m$ for all m , i.e. $(fn_i)_{i \in I}$ is a cocompatible family; therefore there exists t such that $fn_i = tcn_i$ for all i ; then $f = tc$ and $\ker c \leq \ker f$. Finally, let (m,p) be a regular decomposition of c . Since $\ker c = \ker p$ contains all $\mathrm{Im}(n_i \times n_j D_m)$, we conclude as above that $p = tc$ for some t ; it follows that $tm = 1$ and the monomorphism m is in fact a coretraction.

However, $uc = vc$ implies $ucn_i = vcn_i$ for all i and $u = v$, so that c is an epimorphism.

Conversely, assume that c satisfies the conditions in the statement. Then, first, $cn_i = cn_jD_m$ for all $m \in I$, so that $(cn_i)_{i\in I}$ is a cocompatible family. Any other cocompatible family $(f_i)_{i\in I}$ will induce a morphism f from the coproduct, with $f_i = fn_i$ for all i ; by cocompatibility we see that $fn_i = fn_jD_m$ for all m , so that $\ker f$ contains all $\mathrm{Im}(n_i \times n_jD_m)$; hence $\ker f \geq \ker c$, and it follows from 5.4 that $f = tc$ for some t . Then $f_i = tcn_i$ and this factorization is clearly unique, which shows that $(C,(cn_i)_{i\in I})$ is indeed a colimit of $\mathfrak{D}$ and completes the proof.

A similar, but simpler, result exists for coequalizers:

Proposition 6.5. $c \in \mathrm{Coequ}(f,g)$ if and only if c is a regular epimorphism and $\ker c$ is the least congruence containing $\mathrm{Im}(f \times g)$.

Proof. Easy enough, using 6.3, 5.4.

In general we also have a connection between colimits and unions:

Proposition 6.6. Let $\mathfrak{D} : I \longrightarrow \mathcal{C}$ be a diagram with a colimit $(C,(c_i)_{i\in I})$; let $(A,(a_i)_{i\in I})$ be a cocompatible family inducing $a : C \longrightarrow A$. Then $\mathrm{Im}\, a = \bigvee_{i\in I} \mathrm{Im}\, a_i$. In particular $\bigvee_{i\in I} \mathrm{Im}\, c_i = 1$.

Proof. For all i , $a_i = ac_i$, so that $\mathrm{Im}\, a_i \leq \mathrm{Im}\, a$. Conversely let m be a monomorphism such that $\mathrm{Im}\, a_i \leq \mathrm{Im}\, m$ for all i . Then for each i we have $a_i = mt_i$ for some t_i , and since $(a_i)_{i\in I}$ is a cocompatible family and m is a monomorphism we see that $(t_i)_{i\in I}$ is cocompatible. Hence there exists a morphism t with $t_i = tc_i$ for all i , and since $ac_i = mt_i = mtc_i$ for all i we have $a = mt$ and $\mathrm{Im}\, a \geq \mathrm{Im}\, m$. Therefore $\mathrm{Im}\, a = \bigvee_{i\in I} \mathrm{Im}\, a_i$.

4. From these results we see that in a regular category there

are implications between the existence of certain limits and colimits.

Proposition 6.7. A well-powered regular category with coproducts has intersections.

Proof. By well-powered-ness we can produce for each object A a partially ordered set A_s which we could call set of all subobjects of A . By 6.6, A_s is a [small] complete $\vee$-semilattice with a greatest element 1 , hence is a complete lattice. Thus we have intersections.

Intersections of congruences can also be obtained as follows. If we assume that $\mathcal{C}$ is complete, then we always have an intersection for any family of congruences, and the resulting relation is a congruence, by 6.1 (for $\bigwedge_{i\in I} \ker f_i$ is then equal to the congruence $\ker \bigtimes_{i\in I} f_i$).

More interesting is the following result:

Proposition 6.8. Let $\mathcal{C}$ be a regular category in which Lawvere's condition holds. Further assume either that $\mathcal{C}$ is regularly co-well-powered and has intersections, or that $\mathcal{C}$ has unions and that inverse images preserve directed unions. Then, for each relation $\alpha : A \longrightarrow A$ there exists a least congruence on A containing α .

Proof. First let $(\gamma_i)_{i\in I}$ be any family of congruences having an intersection $\gamma = \bigwedge_{i\in I} \gamma_i$. We have $\epsilon \leq \gamma$ since each γ_i is reflexive; also $\gamma^{-1} = (\bigwedge_{i\in I}\gamma_i)^{-1} = \bigwedge_{i\in I} \gamma_i^{-1} = \gamma$ since each γ_i is symmetric; and $\gamma\gamma \leq \gamma_i\gamma_i = \gamma_i$ for each i , so that $\gamma\gamma \leq \gamma$ and γ is transitive; therefore γ is a congruence. Thus, under Lawvere's condition, every existing intersection of congruences is a congruence. Under the first set of further assumptions it now suffices to take the intersection of all congruences on A that contain α.

Under the second set of assumptions, we first let $\beta = \epsilon \vee \alpha \vee \alpha^{-1}$.

Note that $\alpha \leq \beta$ and that β is reflexive and symmetric. We now define β^n by successive compositions: $\beta^{n+1} = \beta^n\beta$ and let $\gamma = \bigvee_{i\in I} \beta^n$; we note that $\beta^n \leq \beta^{n+1}$, since β is reflexive, so that this is a directed union. We claim that γ is the least congruence containing α. Since every congruence which contains α also contains β, and all β^n, it suffices to show that γ is a congruence. It is clear that γ is reflexive, and symmetric (since by induction, all β^n are symmetric). For the transitivity, we use the assumption that inverse images preserve directed unions.[This condition will be called (C_3') in next chapter and the proof of the following facts can be found there in detail.] When applied to inverse images under monomorphisms, it means that intersection with a fixed subobject distributes directed unions. By directedness the same is true for finite intersections in general. Thus we see from Puppe's formula that the composition of relations also distributes directed unions. It follows that $\gamma\gamma = \bigvee_{m,n>0} \beta^m\beta^n = \gamma$, and γ is transitive, which completes the proof.

Corollary 6.9. Let $\mathcal{C}$ be a regular category satisfying either of the following conditions: i) $\mathcal{C}$ is complete and regularly co-well-powered; ii) $\mathcal{C}$ has intersections, is regularly co-well-powered and satisfies Lawvere's condition; iii) $\mathcal{C}$ has unions, satisfies Lawvere's condition and inverse images in $\mathcal{C}$ preserve directed unions. Then $\mathcal{C}$ has coequalizers; if $\mathcal{C}$ has coproducts(finite coproducts), then $\mathcal{C}$ is cocomplete (finitely cocomplete).

Proof. The conclusion of 6.8 will hold in either case and then it follows from 6.5 that $\mathcal{C}$ has coequalizers.

Synopsis of definitions and formulae

$1 \; , \; 1_A$: identity morphism, also greatest subobject

$\underline{x} \wedge \underline{y} \; , \; \bigvee_{i \in I} \underline{x}_i$: l.u.b. of families of subobjects

$\underline{x} \wedge \underline{y} \; , \; \bigwedge_{i \in I} \underline{x}_i$: g.l.b. (intersections) of subobjects

Im m : subobject containing the monomorphism m

Im f : Image of f (= Im m , if (m,p) is a regular decomposition of f.

$f^s \underline{x}$: inverse image of subobject $\underline{x}$ under f

$f_s \underline{x}$: direct image of subobject $\underline{x}$ under f

$$\begin{cases} (1_A)^s \underline{x} = \underline{x} \; , \quad (fg)^s \underline{x} = g^s f^s \underline{x} \; , \quad f^s 1 = 1 \\ f^s(\bigwedge_{i \in I} \underline{x}_i) = \bigwedge_{i \in I} f^s \underline{x}_i \; , \quad \underline{x} \leq \underline{y} \Rightarrow f^s \underline{x} \leq f^s \underline{y} \end{cases}$$

$$\begin{cases} (1_A)_s \underline{x} = \underline{x} \; , \quad (fg)_s \underline{x} = f_s g_s \underline{x} \\ f_s 1 = \mathrm{Im}\, f \; , \quad f_s \, \mathrm{Im}\, g = \mathrm{Im}\, fg \leq \mathrm{Im}\, f \\ f_s(\bigvee_{i \in I} \underline{x}_i) = \bigvee_{i \in I} f_s \underline{x}_i \; , \quad \underline{x} \leq \underline{y} \Rightarrow f_s \underline{x} \leq f_s \underline{y} \end{cases}$$

$$\begin{cases} \underline{x} \leq f^s \underline{y} \iff f_s \underline{x} \leq \underline{y} \\ f_s f^s \underline{y} \leq \underline{y} \; , \quad f^s f_s \underline{x} \geq \underline{x} \\ f_s f^s f_s = f_s \; , \quad f^s f_s f^s = f^s \\ f_s f^s \underline{x} = \underline{x} \wedge \mathrm{Im}\, f \\ f \;\; \text{reg.epi} \Rightarrow f_s f^s \underline{x} = \underline{x} \\ f \;\; \text{mono} \Rightarrow f^s f_s \underline{y} = \underline{y} \end{cases}$$

α, β, γ : relations

$\epsilon \; , \; \epsilon_A$: ("equality" , "diagonal") least congruence on A

$\upsilon \; , \; \upsilon_A$: $= 1_{\tilde{A}}$, greatest congruence on A

$\mathrm{Im}\, \alpha = \mathrm{Im}\, a$ if $\alpha = \mathrm{Im}(a \times b)$

α^{-1} : inverse of α ; $\alpha^{-1} = \mathrm{Im}(b \times a)$ when $\alpha = \mathrm{Im}(a \times b)$

$\tilde{A} = A \sqcap A$, $\tilde{f} = f \sqcap f$: tilda functor, preserves limits and regular decompositions

ker f : congruence induced by f , $= f^{-1} f = \mathrm{Im}(x \times y)$ where $fx = fy$

is a pullback

$$\begin{cases}\epsilon^{-1} = \epsilon \ , \ \alpha \le \beta \Longrightarrow \alpha^{-1} \le \beta^{-1} \ , \ (\alpha^{-1})^{-1} = \alpha \\ (\bigvee_{i\in I} \alpha_i)^{-1} = \bigvee_{i\in I} \alpha_i^{-1} \ , \ (\bigwedge_{i\in I} \alpha_i)^{-1} = \bigwedge_{i\in I} \alpha_i^{-1}\end{cases}$$

$$\left\{\mathrm{Im}(\bigvee_{i\in I} \alpha_i) = \bigvee_{i\in I} \mathrm{Im}\, \alpha_i \ , \ \alpha \le \beta \Longrightarrow \mathrm{Im}\, \alpha \le \mathrm{Im}\, \beta\right.$$

$$\begin{cases}\beta\alpha = r_s(p^s\alpha \wedge q^s\beta) \quad \text{(Puppe's formula)} \\ \epsilon\alpha = \alpha \ , \ \beta\epsilon = \beta \ , \ (\alpha\beta)\gamma = \alpha(\beta\gamma) \\ \alpha \le \alpha' \ , \ \beta \le \beta' \Longrightarrow \alpha\beta \le \alpha'\beta' \ ; \ (\alpha\beta)^{-1} = \beta^{-1}\alpha^{-1} \\ \alpha = \mathrm{Im}(a \times b) \Longrightarrow \alpha = ba^{-1}\end{cases}$$

$\alpha_s\underline{x}$: direct image of subobject $\underline{x}$ under relation α

$\alpha^s\underline{x}$: inverse image of subobject $\underline{x}$ under relation α

$$\begin{cases}\epsilon_s\underline{x} = \underline{x} \ , \ (\alpha\beta)_s\underline{x} = \alpha_s\beta_s\underline{x} \ , \ \alpha_s 1 = \mathrm{Im}\, \alpha \\ \underline{x} \le \underline{y} \Longrightarrow \alpha_s\underline{x} \le \alpha_s\underline{y} \ , \ \alpha \le \beta \Longrightarrow \alpha_s\underline{x} \le \beta_s\underline{x} \\ \mathrm{Im}\, \alpha\beta = \alpha_s \, \mathrm{Im}\, \beta \ , \ \alpha = \mathrm{Im}(a \times b) \Longrightarrow \alpha_s = b_s a^s\end{cases}$$

$$\begin{cases}\alpha^s = (\alpha^{-1})_s \ , \ \alpha_s = (\alpha^{-1})^s \\ \epsilon^s\underline{x} = \underline{x} \ , \ (\alpha\beta)^s\underline{x} = \beta^s\alpha^s\underline{x} \ , \ \alpha = \mathrm{Im}(a \times b) \Longrightarrow \alpha^s = a_s b^s \\ \underline{x} \le \underline{y} \Longrightarrow \alpha^s\underline{x} \le \alpha^s\underline{y} \ , \ \alpha \le \beta \Longrightarrow \alpha^s\underline{x} \le \beta^s\underline{x}\end{cases}$$

$$\begin{cases}ff^{-1} \le \epsilon \ , \ f^{-1}f \ge \epsilon \\ f \ \text{reg. epi} \Longleftrightarrow ff^{-1} = \epsilon \\ f \ \text{mono} \Longleftrightarrow f^{-1}f = \epsilon\end{cases}$$

$f_r\alpha$: direct image of relation α under morphism f

$f^r\alpha$: inverse image of relation α under morphism f

$$\begin{cases}f_r\alpha = f\alpha f^{-1} = \tilde{f}_s\alpha \ , \ f^r\alpha = f^{-1}\alpha f = \tilde{f}^s\alpha \\ f^r \ker g = \ker gf \ge \ker f \\ f^r\epsilon = \ker f \ , \ f_r \ker f \le \epsilon\end{cases}$$

II. DIRECTED COLIMITS IN REGULAR CATEGORIES

Our first result gives necessary and sufficient conditions, of an elementary nature, that directed colimits in a given cocomplete regular category be exact. In the abelian case, Grothendieck showed that the subobject condition

A.B.5: $\underline{x} \wedge (\bigvee_{i \in I} \underline{y}_i) = \bigvee_{i \in I} (\underline{x} \wedge \underline{y}_i)$ whenever $(y_i)_{i \in I}$ is directed

is necessary and sufficient [15],[31] . In the case of a regular category, the necessary and sufficient condition comes in three parts:

(C_3') Inverse images preserve directed unions of subobjects;

(C_3'') A directed union of congruences is a congruence;

(C_3''') If $(X_i)_{i \in I}$ is the family of objects of a monic direct system [= in which all morphisms $X_i \longrightarrow X_j$ $(i \leq j)$ are monomorphisms], there exists a family of monomorphisms $X_i \longrightarrow C$ (not necessarily a cocompatible family).

In the abelian case, (C_3'') and (C_3''') evaporate. The remaining condition (C_3') is still stronger that A.B.5 (though no harder to verify on the examples): the extra strength is used in the proof to manipulate relations (which are not needed in the abelian case). It implies (C_3'') when Lawvere's condition on congruences holds; and (C_3''') holds whenever coproduct injections are monomorphisms, so these are fairly mild conditions.

The proof occupies most of this part. It is somewhat technical; also, unlike what happens in the abelian case, preservation of finite limits has to be established, and even though it implies preservation of monomorphisms, the latter has to be shown first anyway. In the we obtain additional results showing that when directed colimits are exact they show additional good behavior: for instance, Gray's condition $\mathfrak{J}_2$ holds.

A cocomplete regular category in which directed colimits are exact is called a C_3 regular category. It is called C_4 if in addition it is complete and satisfies Gray's condition $\mathfrak{J}_1$ [10],[31], which is the same as Grothendieck's condition A.B.6 [15]. In the last section we show that in a C_4 regular category any product of directed colimits can be rewritten as a directed colimit of products, provided that all direct systems under consideration are monic. The last restriction can be lifted if furthermore the category is [regularly] C_1^*, i.e. any product of regular epimorphisms is a regular epimorphism. Of course all these conditions hold in a C_4, C_1^* abelian category, as well as in any finitary variety.

All these results are taken from [14]. References such as I.x.y refer to result x.y in part I above; we use the same conventions as in that part. Throughout, I will also denote a directed preordered set. A direct system $\mathfrak{X}$ over I is a functor of domain I, and we write X_i for $\mathfrak{X}(i)$ $(i \in I)$ and $x_{ij} : X_i \longrightarrow X_j$ $(i \leq j)$, the objects and morphisms in the system. We denote $\varinjlim \mathfrak{X} = \varinjlim_{i \in I} X_i$ by X and $X_i \longrightarrow X$ by x_i. Similar conventions apply to direct systems $\mathfrak{Y}, \mathfrak{Z}$, etc.

It is suggested that the reader be well-acquainted with the techniques developed in the first part before reading the proofs which follow.

1. THE MAIN THEOREM: DIRECT PART.

1. In this part we let G be C_3 regular category, i.e. we assume that G is cocomplete and that directed colimits are exact. Note that for each directed preordered set I the functor category $[I,G]$ is regular, by I.2.1, with pointwise decompositions, finite limits and colimits, so that in particular it makes sense to say that the colimit functor is exact. We shall show that (C_3'), (C_3''), (C_3''') hold in G.

2. Let $f : A \longrightarrow B \in G$ and $(\underline{x}_i)_{i\in I}$ be a directed family of subobjects of B. We may define $i \leq j$ if and only if $\underline{x}_i \leq \underline{x}_j$ and then I becomes a directed preordered set; a direct system $\mathfrak{X} : I \longrightarrow G$ is then constructed as follows. Since I is a set we can select for each $i \in I$ a monomorphism m_i with $\operatorname{Im} m_i = \underline{x}_i$. Let X_i be the domain of m_i. If $i \leq j$, then $\operatorname{Im} m_i \leq \operatorname{Im} m_j$ and $m_i = m_j x_{ij}$ for some unique $x_{ij} : X_i \longrightarrow X_j$; by the uniqueness, it is clear that we now have a direct system. In addition, we have a [pointwise] monomorphism $\mathfrak{m} = (m_i)_{i\in I} : \mathfrak{X} \longrightarrow B$. Since G is C_3, the induced morphism $m : X \longrightarrow B$ is a monomorphism; by I.6.6, $\operatorname{Im} m = \bigvee_{i\in I} \operatorname{Im} m_i = \bigvee_{i\in I} \underline{x}_i$.

For each i, $f^s \underline{x}_i = \operatorname{Im} n_i$, where $f n_i = m_i g_i$ is a pullback. As above, there is a direct system $\mathfrak{Y} : I \longrightarrow G$ with a [pointwise] monomorphism $\mathfrak{n} = (n_i)_{i\in I} : \mathfrak{Y} \longrightarrow A$ (note that $i \leq j$ implies $f^s \underline{x}_i \leq$ $\leq f^s \underline{x}_j$); the induced morphism $n : Y \longrightarrow A$ is a monomorphism and satisfies $\operatorname{Im} n = \bigvee_{i\in I} f^s \underline{x}_i$.

We also have a morphism $(g_i)_{i\in I} : \mathfrak{Y} \longrightarrow \mathfrak{X}$. Since $f\mathfrak{n} = \mathfrak{m}\mathfrak{g}$ is a pullback and G is C_3, the colimit square $fn = mg$ is also a pullback; hence $\bigvee_{i\in I} f^s \underline{x}_i = \operatorname{Im} n = f^s \operatorname{Im} m = f^s(\bigvee_{i\in I} \underline{x}_i)$. Therefore (C_3') holds.

3. The verification of (C_3'') is similar. Let $(\alpha_i)_{i\in I}$ be a directed family of congruences on $A \in G$. Write $\alpha_i = \operatorname{Im}(x_i \times y_i) = \ker p_i$,

where p_i is a regular epimorphism and $p_i x_i = p_i y_i$ is a pullback, so that $x_i \times y_i$ is a monomorphism. A direct system $\mathfrak{X}$ is constructed as above, so that $(x_i \times y_i)_{i \in I}$ is a monomorphism $\mathfrak{X} \longrightarrow A \sqcap A$. In addition, $i \leq j$ implies $\ker p_i = \alpha_i \leq \alpha_j = \ker p_j$, so that by the induced homomorphism theorem (I.5.4) we have $p_j = b_{ij} p_i$ for some unique b_{ij}. From this we obtain a direct system $\mathfrak{B}$ such that $(p_i)_{i \in I}$ is a regular epimorphism $A \longrightarrow \mathfrak{B}$. Since $p_i x_i = p_i y_i$ is a pullback, we obtain at the colimit a pullback $px = py$. There $x \times y$ is also the colimit of $(x_i \times y_i)_{i \in I}$ and by I.6.6 $\operatorname{Im}(x \times y) = \bigvee_{i \in I} \operatorname{Im}(x_i \times y_i) = \bigvee_{i \in I} \alpha_i$. It follows that $\bigvee_{i \in I} \alpha_i = \ker p$ is a congruence.

4. The verification of (C_3''') is less straightforward. It follows from the slightly more general result:

<u>Lemma</u> 1.1. Let $\mathfrak{C}$ be a finitely complete category which has directed colimits that preserve monomorphisms. If $\mathfrak{X}$ is a monic direct system in $\mathfrak{C}$ over I, then every morphism $x_i : X_i \longrightarrow X$ is a monomorphism.

<u>Proof</u>. The proof is immediate if I happens to be a directed $\wedge$-semilattice. In that case there is for each $i \in I$ a direct system $\mathfrak{Y} : I \longrightarrow \mathfrak{C}$, defined by: $Y_j = X_{i \wedge j}$, $y_{jk} = x_{i \wedge j, i \wedge k}$ $(j \leq k)$. Also there is a monomorphism $(x_{i \wedge j, j})_{j \in I} : \mathfrak{Y} \longrightarrow \mathfrak{X}$; we claim its colimit is precisely x_i. First note that (up to isomorphism) $Y = X_i$, with $y_j = x_{i \wedge j, i}$; then, for all $j \in I$, $x_i \, y_j = x_i \, x_{i \wedge j, i} = x_{i \wedge j} = x_j \, x_{i \wedge j, j}$ – which proves the claim. By the hypothesis on $\mathfrak{C}$, x_i is then a monomorphism.

If now I is arbitrary, then we come back to the case of a directed $\wedge$-semilattice as follows. First we find the semilattice. For each $k \in I$, let S_k be the set of all intersections of finitely many subobjects of X_k of the form $\operatorname{Im} x_{ik}$ $(i \leq k)$. Note that S_k is an

$\wedge$-semilattice. If $k \leq \ell$ in I, a map $s_{k\ell} : S_k \longrightarrow S_\ell$ is defined as follows. Since $x_{k\ell} : X_k \longrightarrow X_\ell$ is a monomorphism, we do not need regular decompositions to have direct images under $x_{k\ell}$; in addition, direct images under $x_{k\ell}$ preserve intersections (for finite intersections: if $mn' = nm'$ is a pullback, then so is $(x_{k\ell}m)n' = (x_{k\ell}n)m'$). Hence $(x_{k\ell})_s$ restricts to a mapping $s_{k\ell} : S_k \longrightarrow S_\ell$, which is in fact an injective homomorphism of $\wedge$-semilattices. Furthermore, x_{kk} is the identity, hence so is s_{kk} ; if $k \leq \ell \leq m$ in I, then $x_{km} = x_{\ell m}x_{k\ell}$, hence $(x_{km})_s = (x_{\ell m})_s(x_{k\ell})_s$ and $s_{km} = s_{\ell m}s_{k\ell}$; in other words, we now have a direct system of $\wedge$-semilattices; this yields an $\wedge$-semilattice $S = \varinjlim S_k$, which comes with injective homomorphisms $s_k : S_k \longrightarrow S$ such that $S = \bigcup_{k \in I} s_k(S_k)$.

An order-preserving map $i \longmapsto \hat{i}$, $I \longrightarrow S$, is defined by: $\hat{i} = s_i(\mathrm{Im}\ x_{ii}) = s_i(1)$; it is order-preserving since $i \leq j$ implies $\hat{i} = s_j(s_{ij}(\mathrm{Im}\ x_{ii})) = s_j(\mathrm{Im}\ x_{ij}) \leq s_j(\mathrm{Im}\ x_{jj})$ since each s_j is order-preserving. The image $\hat{I} = [\hat{i} ; i \in I]$ is cofinal in S since for each $u \in S$ we have $u \in s_k(S_k)$ for some k and therefore $u \leq \hat{k}$. It follows that S is directed.

For each $u \in S$, select $k \in I$ with $u \in s_k(S_k)$ and a monomorphism $y_{uk} : Y_u \longrightarrow X_k$ such that $u = s_k(\mathrm{Im}\ y_{uk})$ [it is easy to see that a different choice only replaces Y_u by an isomorphic object]. Now assume that $u \leq v$ in S and that $y_{v\ell} : Y_v \longrightarrow X_\ell$ has been selected for v (so that $v = s_\ell(\mathrm{Im}\ y_{v\ell})$) . Since I is directed, we have $k \leq m$, $\ell \leq m$ for some $m \in I$; then also

$$s_m(\mathrm{Im}\ x_{km}y_{uk}) = s_m(s_{km}(\mathrm{Im}\ y_{uk})) = s_k(\mathrm{Im}\ y_{uk}) = u$$

and similarly $v = s_m(\mathrm{Im}\ x_{\ell m}y_{v\ell})$. Now s_m is an injective homomorphism and therefore reflects order [$s_m(a) \leq s_m(b)$ implies $s_m(a) = s_m(a) \wedge s_m(b) = s_m(a \wedge b)$ and $a = a \wedge b \leq b$]; hence, $u \leq v$ implies $\mathrm{Im}\ x_{km}y_{uk} \leq \mathrm{Im}\ x_{\ell m}y_{v\ell}$ and there exists a unique $y_{uv} : Y_u \longrightarrow Y_v$ such that $x_{km}y_{uk} = x_{\ell m}y_{v\ell}y_{uv}$. Note that y_{uv} is a monomorphism. If furthermore $m \leq n$ in I, then

$$x_{kn}y_{uk} = x_{mn}x_{km}y_{uk} = x_{mn}x_{\ell m}y_{v\ell}y_{uv} = x_{\ell n}y_{v\ell}y_{uv} ;$$

by the uniqueness, y_{uv} would be the same if we had started from $n \geq k$, instead of m. Since I is directed, it follows that y_{uv} does not depend on the choice of m (as long as m is large enough). If now $u = v$, then $k = \ell$, $y_{uk} = y_{v\ell}$ and since $x_{km}y_{uk} = x_{\ell m}y_{v\ell}1$ it follows that $y_{uu} = 1$. If $u \leq v \leq w$ in S, and we have selected $y_{wm} : Y_w \longrightarrow X_m$ for w and chosen n large enough, then

$$x_{mn}y_{wm}y_{uw} = x_{km}y_{uk} = x_{\ell n}y_{v\ell}y_{uv} = x_{mn}y_{wm}y_{vw}y_{uv}$$

shows that $y_{uw} = y_{vw}y_{uv}$. Therefore we have another monic direct system $\psi : S \longrightarrow G$.

From ψ we obtain a direct system $\psi' : I \longrightarrow G$, defined by $Y'_i = Y_{\hat{\imath}}$, $y'_{ij} = y_{\hat{\imath}\hat{\jmath}}$ $(i \leq j)$; we claim that it is isomorphic to χ. To see this, take $i \in I$. To $\hat{\imath} \in S$ we have associated $y_{\hat{\imath}k} : Y_{\hat{\imath}} \longrightarrow X_k$ (with $\hat{\imath} = s_k(\mathrm{Im}\, y_{\hat{\imath}k})$); we cannot assume that $k = i$ since it may happen that $\hat{\imath} = \hat{\jmath}$ with $i \neq j$; but we may assume that $i \leq k$, for then we have seen that $\hat{\imath} \in s_k(S_k)$. Then

$$s_k(\mathrm{Im}\, y_{\hat{\imath}k}) = \hat{\imath} = s_i(\mathrm{Im}\, x_{ii}) = s_k(s_{ik}(\mathrm{Im}\, x_{ii})) = s_k(\mathrm{Im}\, x_{ik})$$

shows that $\mathrm{Im}\, y_{\hat{\imath}k} = \mathrm{Im}\, x_{ik}$ and therefore there is an isomorphism $a_i : Y_{\hat{\imath}} \longrightarrow X_i$ such that $y_{\hat{\imath}k} = x_{ik}a_i$. If $i \leq j$ in I and we have selected $y_{\hat{\jmath}\ell} : Y_{\hat{\jmath}} \longrightarrow X_\ell$ for $\hat{\jmath}$ and chosen m large enough, then

$$x_{jm}x_{ij}a_i = x_{km}x_{ik}a_i = x_{km}y_{\hat{\imath}k} = x_{\ell m}y_{\hat{\jmath}\ell}y_{\hat{\imath}\hat{\jmath}} = x_{\ell m}x_{j\ell}a_jy_{\hat{\imath}\hat{\jmath}} = x_{jm}a_jy_{\hat{\imath}\hat{\jmath}}$$

(since $\hat{\imath} \leq \hat{\jmath}$), so that $x_{ij}a_i = a_jy_{\hat{\imath}\hat{\jmath}}$. Therefore $(a_i)_{i\in I} : \psi' \longrightarrow \chi$ is an isomorphism.

Since $\hat{I}$ is cofinal in S, it is clear that the obvious morphism $\psi' \longrightarrow \psi$ induces an isomorphism at the colimits. Now S is a directed

$\wedge$-semilattice, and it follows from the first part of the proof that $y_f : Y_f \longrightarrow Y$ is a monomorphism. Using the isomorphisms $X \cong Y' \cong Y$ we conclude that $x_1 : X_1 \longrightarrow X$ is a monomorphism, q.e.d.

5. We have now proved the direct part of the main theorem in this part, namely:

Theorem 1.2. A cocomplete regular category is C_3 if and only if it satisfies (C_3'), (C_3'') and (C_3''') .

2. CONVERSE: PRESERVATION OF MONOMORPHISMS.

1. We now assume that $\mathcal{C}$ is a cocomplete regular category which satisfies (C_3'), (C_3'') and (C_3''') and begin with a few easy consequences of (C_3') .

Proposition 2.1. Under (C_3'), finite intersections of subobjects and composition of relations distribute directed unions.

Proof. If first m is a monomorphism, then it follows from the definitions (or from I.3.7) that $m_s m^s \underline{x} = \operatorname{Im} m \wedge \underline{x}$ for all $\underline{x}$. If now $(\underline{x}_i)_{i\in I}$ is a directed family of subobjects of the codomain of m, then, by (C_3') and I.3.3,

$$\operatorname{Im} m \wedge (\bigvee_{i\in I} \underline{x}_i) = m_s(m^s(\bigvee_{i\in I} \underline{x}_i)) = m_s(\bigvee_{i\in I} m^s\underline{x}_i) = \bigvee_{i\in I} m_s m^s \underline{x}_i = \bigvee_{i\in I}(\operatorname{Im} m \wedge \underline{x}_i) .$$

This shows that intersections by a fixed subobject distributes directed unions. If now $(\underline{y}_j)_{j\in J}$ is another directed family, then, applying this to each $\underline{x}_i$ and then to $\bigvee_{j\in J} \underline{y}_j$, we obtain:

$$\bigvee_{\substack{i\in I\\ j\in J}} (\underline{x}_i \wedge \underline{y}_j) = \bigvee_{i\in I}(\bigvee_{j\in J}(\underline{x}_i \wedge \underline{y}_j)) = \bigvee_{i\in I}(\underline{x}_i \wedge (\bigvee_{j\in J} \underline{y}_j)) = (\bigvee_{i\in I} \underline{x}_i) \wedge (\bigvee_{j\in J} \underline{y}_j),$$

which proves the first assertion. The second assertion is them immediate on Puppe's formula.

Corollary 2.2. If Lawvere's condition (L) holds, then (C_3') implies (C_3'').

Proof. Let $(\alpha_i)_{i\in I}$ be a directed family of congruences [with $I \neq \emptyset$] and $\alpha = \bigvee_{i\in I} \alpha_i$. It is clear that α is reflexive and symmetric; in view of (L) it suffices to prove that α is transitive (i.e. $\alpha\alpha \leq \alpha$). By 2.1, $\alpha\alpha = \bigvee_{j,k\in I} \alpha_j\alpha_k$. Now $\bigvee_{i\in I} \alpha_i\alpha_i \leq \bigvee_{j,k\in I} \alpha_j\alpha_k$ since the index set on the left is smaller; but the converse inequality holds since $(\alpha_i)_{i\in I}$ is directed. The transitivity of α then follows from that of each α_i.

It follows from 2.2 and I.5.2 that (C_3'') is superfluous in case G is abelian.

2. We now start a closer study of direct systems.

Lemma 2.3. Let $a_i : X_i \longrightarrow A$ $(i \in I)$ be a cocompatible family for the direct system $X\,[:I \longrightarrow G]$, inducing $a : X \longrightarrow A$. Then $a = \bigvee_{i\in I} a_i x_i^{-1}$.

Proof. First, $i \leq j$ implies $a_i x_i^{-1} = a_j x_{ij} x_{ij}^{-1} x_j^{-1} \leq a_j x_j^{-1}$; hence $(a_i x_i^{-1})_{i\in I}$ is a directed family of relations. Hence

$$
\begin{aligned}
(\bigvee_{i\in I} a_i x_i^{-1})(\bigvee_{j\in J} a_j x_j^{-1})^{-1} &= \bigvee_{i,j\in I} a_i x_i^{-1} x_j a_j^{-1} \leq \\
&\leq \bigvee_{k\in I} a_k x_k^{-1} x_k a_k^{-1} = \bigvee_{k\in I} a\, x_k x_k^{-1} x_k x_k^{-1} a^{-1} = \\
&= \bigvee_{k\in I} a\, x_k x_k^{-1} a^{-1} = \bigvee_{k\in I} (a x_k)(a x_k)^{-1} \leq \varepsilon .
\end{aligned}
$$

Since also $(\bigvee_{i\in I} a_i x_i^{-1})^{s}1 = \bigvee_{i\in I} (x_i)_s a^{s}1 = \bigvee_{i\in I} \operatorname{Im} x_i = 1$ by I.6.6, it follows from I.4.12 that $b = \bigvee_{i\in I} a_i x_i^{-1}$ is a morphism. To show that $b = a$, we note that, since I is directed, $b = \bigvee_{j\geq i} a_j x_j^{-1}$, for every $i \in I$; hence

$$bx_i = (\bigvee_{j\geq i} a_j x_j^{-1})x_i = \bigvee_{j\geq i} a_j x_j^{-1} x_i = \bigvee_{j\geq i} a_j x_j^{-1} x_j x_{ij} \geq$$

$$\geq \bigvee_{j\geq i} a_j x_{ij} = \bigvee_{j\geq i} a_i = a_i = ax_i \quad ;$$

since bx_i and ax_i are morphisms, this implies $bx_i = ax_i$; it holds for every i, hence $b = a$.

Corollary 2.4. If in 2.3 each a_i is a monomorphism, then a is a monomorphism.

Proof. $\ker a = (\bigvee_{i\in I} a_i x_i^{-1})^{-1}(\bigvee_{j\in I} a_j x_j^{-1}) \leq \bigvee_{k\in I} x_k a_k^{-1} a_k x_k^{-1} \leq \varepsilon$.

3. We now establish progressively stronger results. The next one already uses the full strength of the hypothesis.

Lemma 2.5. If $\mathfrak{X}$ is a monic direct system, then each x_i is a monomorphism.

Proof. Let $C = \coprod_{i\in I} X_i$ be the coproduct, with injections $m_i : X_i \longrightarrow C$; by (C_3'''), each m_i is a monomorphism. It follows from I.6.4 that there is a regular epimorphism $c : C \longrightarrow X$ such that $x_i = cm_i$ for all i, and that $\ker c$ is the least congruence on C that contains every $\mathrm{Im}(m_i \times m_j x_{ij})$ with $i \leq j$.

Let $\mathfrak{F}$ be the set of all finite subsets of $\{(i,j) \in I \sqcap I ; i \leq j\}$ [= of the preorder relation on I]. For each $F \in \mathfrak{F}$, the subdiagram of $\mathfrak{X}$ consisting of all X_i with only those x_{ij} with $(i,j) \in F$ has a colimit in $\mathfrak{C}$; again by I.6.4, there exists a least congruence α_F on C containing all $\mathrm{Im}(m_i \times m_j x_{ij})$ with $(i,j) \in F$. From that property it is clear that $F \subseteq G$ implies $\alpha_F \leq \alpha_G$, so that $(\alpha_F)_{F\in\mathfrak{F}}$ is a directed family of congruences. By (C_3''), $\alpha = \bigvee_{F\in\mathfrak{F}} \alpha_F$ is a congruence. Now the "least" property of α_F implies that $\alpha_F \leq \ker c$ for every F, so that $\alpha \leq \ker c$; the converse inequality follows from the similar property of $\ker c$, since α is a congruence. We conclude that $\ker c = \bigvee_{F\in\mathfrak{F}} \alpha_F$. Now if we can prove that $m_i^r \alpha_F \leq \varepsilon$ for all i and F, it will follows that

$$\ker x_i = \ker cm_i = m_i^r \ker c = m_i^r(\bigvee_{F\in\mathfrak{F}} \alpha_F) = \bigvee_{F\in\mathfrak{F}} m_i^r \alpha_F \leq \varepsilon ,$$

and the lemma will be proved.

For each $i \in I$, $F \in \mathfrak{F}$, there is a $t \in I$ with $i \leq t$ and $j \leq t$, $k \leq t$ for all $(j,k) \in F$ (since F is finite). Consider the diagram:

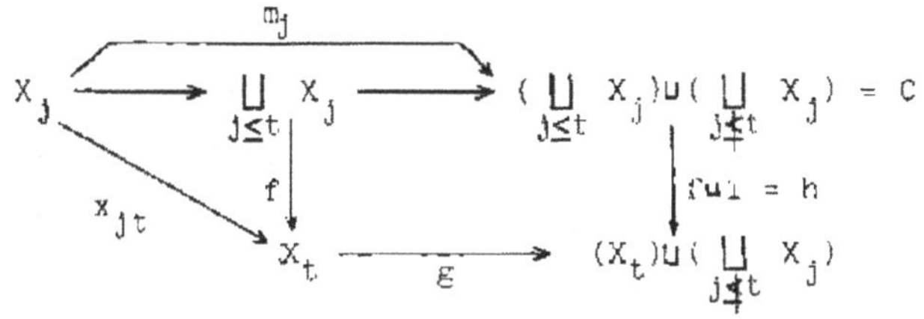

where g as well as all unnamed maps are coproduct injections, and f is induced by all x_{jt} , $j \leq t$, so that the diagram commutes for every $j \leq t$.

If $(j,k) \in F$, then $j \leq t$, $k \leq t$ and we see on the diagram that

$$hm_j = gx_{jt} = gx_{kt}x_{jk} = hm_k x_{jk} \quad ;$$

then it follows from I.6.3 that $\operatorname{Im}(m_j \times m_k x_{jk}) \leq \ker h$; therefore $\alpha_F \leq \ker h$. On the other hand, g is a monomorphism (since m_t is a monomorphism), and so is x_{it} . Hence

$$m_i^r \alpha_F \leq m_i^r \ker h = \ker hm_i = \ker gx_{it} = \varepsilon ,$$

which completes the proof.

The next result gives one of the nice properties of directed colimits in our situation.

Proposition 2.6. For any direct system $\mathfrak{X} : I \longrightarrow \mathfrak{G}$ [where $\mathfrak{G}$ is a C_3 regular category], $\ker x_i = \bigvee_{j\geq i} \ker x_{ij}$ for every $i \in I$.

Proof. It is based on another construction of directed colimits which is somewhat more 'set-like'. First $j \leq k$ implies $\ker x_{ij} \leq \leq \ker x_{jk}x_{ij} = \ker x_{ik}$; it follows that $(\ker x_{ij})_{j\geq i}$ is a directed family of congruences, so that by (C_3'') $\alpha_i = \bigvee_{j\geq i} \ker x_{ij}$ is a congruence for every $i \in I$.

Put $\alpha_i = \ker p_i$, where $p_i : X_i \longrightarrow Y_i$ is a regular epimorphism. If $i \leq j$, then by (C_3'):

$$\ker p_j x_{ij} = \bar{x}_{ij}^s(\bigvee_{k\geq j} \ker x_{jk}) = \bigvee_{k\geq j} \ker x_{ik} = \ker p_i$$

since I is directed; by I.5.4, $p_j x_{ij} = y_{ij} p_i$ for some unique $y_{ij} : Y_i \longrightarrow Y_j$, and y_{ij} is a monomorphism. The uniqueness implies that we now have a monic direct system $\mathfrak{Y} : I \longrightarrow \mathfrak{a}$. We now prove that the morphism $(p_i)_{i\in I} : \mathfrak{X} \longrightarrow \mathfrak{Y}$ induces an isomorphism on the colimits. First it is clear that $(y_i p_i)_{i\in I}$ is a cocompatible family for $\mathfrak{X}$. If $(a_i)_{i\in I}$ is any cocompatible family for $\mathfrak{X}$, then $i \leq j$ implies $\ker x_{ij} \leq \ker a_j x_{ij} = \ker a_i$; therefore $\ker p_i = \alpha_i \leq \ker a_i$; therefore $a_i = b_i p_i$ for some unique b_i ; the uniqueness easily implies that $(b_i)_{i\in I}$ is a cocompatible family for $\mathfrak{Y}$ [equivalently, one may use the induced homomorphism theorem in $[I,\mathfrak{a}]$]. This yields a morphism b unique such that $b_i = by_i$ for all i . We see that $a_i = by_i p_i$ for all i , and the uniqueness of b in this factorization follows from the other uniquenesses. Thus $(y_i p_i)_{i\in I}$ is a colimit of $\mathfrak{X}$, and there is an isomorphism t such that $tx_i = y_i p_i$ for all i .

Now y_i is a monomorphism, by 2.5, so that $\ker x_i = \ker tx_i = = \ker p_i = \bigvee_{j\geq i} \ker x_{ij}$, q.e.d.

4. We now give a lemma which is crucial for the next three proofs of preservation properties.

Lemma 2.7. Let $a_i : A_i \longrightarrow A$ be a family of morphisms such that $(\mathrm{Im}\, a_i)_{i\in I}$ is directed and $\bigvee_{i\in I} \mathrm{Im}\, a_i = 1$. Then $\bigvee_{i\in I} \mathrm{Im}\, \tilde{a}_i = 1$.

Proof. If $p : A \sqcap A \longrightarrow A$, $p_i : A_i \sqcap A \longrightarrow A_i$ are the first projections, then $p(a_i \sqcap 1_A) = a_i p_i$ is a pullback, so that, by I.3.10, $\mathrm{Im}(a_i \sqcap 1_A) = p^s\, \mathrm{Im}\, a_i$. Hence it follows from (C_3') that $\bigvee_{i\in I} \mathrm{Im}(a_i \sqcap 1_A) = 1$. Similarly, $\bigvee_{j\in I} \mathrm{Im}(1_{A_i} \sqcap a_j) = 1$ for each $i \in I$. Since $a_i \sqcap a_j = (a_i \sqcap 1_A)(1_{A_i} \sqcap a_j)$, it follows that

$$\begin{aligned}\bigvee_{i,j\in I} \mathrm{Im}(a_i \sqcap a_j) &= \bigvee_{i\in I}\Big(\bigvee_{j\in I} (a_i \sqcap 1_A)_s\, \mathrm{Im}(1_{A_i} \sqcap a_j)\Big) = \\ &= \bigvee_{i\in I}\Big((a_i \sqcap 1_A)_s\Big(\bigvee_{j\in I} \mathrm{Im}(1_{A_i} \sqcap a_j)\Big)\Big) = \\ &= \bigvee_{i\in I} \mathrm{Im}(a_i \sqcap 1_A) = 1 .\end{aligned}$$

Now if (m_i, p_i) is a regular decomposition of a_i for every i, we have $\mathrm{Im}(a_i \sqcap a_j) = \mathrm{Im}(m_i \sqcap m_j)$ since by I.1.12 finite products preserve decompositions. If $\mathrm{Im}\, a_i \le \mathrm{Im}\, a_k$, $\mathrm{Im}\, a_j \le \mathrm{Im}\, a_k$, then m_i, m_j factor though m_k and therefore $\mathrm{Im}(a_i \sqcap a_j) \le \mathrm{Im}(a_k \sqcap a_k)$. It follows that $(\mathrm{Im}\, \tilde{a}_i)_{i\in I}$ is cofinal in $(\mathrm{Im}(a_i \sqcap a_j))_{i,j\in I}$. The result follows.

We now are in position to prove that directed colimits in $\mathcal{G}$ preserve monomorphisms.

Let $\mathfrak{m} = (m_i)_{i\in I} : \mathfrak{X} \longrightarrow \mathfrak{Y}$ be a [pointwise] monomorphism of direct systems over I, and $m : X \longrightarrow Y$ be induced by $\mathfrak{m}$. By 2.7, $\bigvee_{i\in I} \mathrm{Im}\, \tilde{x}_i = 1$. Hence it follows from 2.6 that:

$$\begin{aligned}\ker m = \ker m \wedge \Big(\bigvee_{i\in I} \mathrm{Im}\, \tilde{x}_i\Big) &= \bigvee_{i\in I}(\ker m \wedge \mathrm{Im}\, \tilde{x}_i) = \\ &= \bigvee_{i\in I} (\tilde{x}_i)_s\, \tilde{x}_i^{\,s} \ker m = \bigvee_{i\in I} (\tilde{x}_i)_s \ker m x_i = \\ &= \bigvee_{i\in I} (\tilde{x}_i)_s \ker y_i m_i = \bigvee_{i\in I} (\tilde{x}_i)_s\, \tilde{m}_i^{\,s} \ker y_i = \\ &= \bigvee_{i\in I} (\tilde{x}_i)_s\, \tilde{m}_i^{\,s}\Big(\bigvee_{j\ge i} \ker y_{ij}\Big) = \bigvee_{j\ge i\in I} (\tilde{x}_i)_s\, \tilde{m}_i^{\,s} \ker y_{ij} = \\ &= \bigvee_{j\ge i\in I} (\tilde{x}_i)_s \ker y_{ij} m_i = \bigvee_{j\ge i\in I} (\tilde{x}_i)_s \ker m_j x_{ij} =\end{aligned}$$

$$= \bigvee_{j \geq i \in I} (\tilde{x}_i)_s \ker x_{ij} = \bigvee_{i \in I} (\tilde{x}_i)_s (\bigvee_{j \geq i} \ker x_{ij}) =$$
$$= \bigvee_{i \in I} (\tilde{x}_i)_s \ker x_i \leq \varepsilon \quad ,$$

which proves that m is a monomorphism.

3. CONVERSE: PRESERVATION OF FINITE LIMITS.

1. Directed colimits already preserve finite colimits (and regular epimorphisms), hence to prove exactness it now suffices to show that they preserve finite limits. One may consider this section as the proof of the converse proper, the previous section (including preservation of monomorphisms) containing only lemmas. We successively prove that directed colimits preserve equalizers, and finite products.

2. Let $\mathcal{E} \xrightarrow{m} \mathfrak{X} \overset{f}{\underset{g}{\rightrightarrows}} \mathfrak{Y}$ be an equalizer diagram in $[I,G]$; well, a cockney equalizer, what; we want to show that the colimit diagram $E \xrightarrow{m} X \overset{f}{\underset{g}{\rightrightarrows}} Y$ is an equalizer diagram (in G). By the previous section we know that m is a monomorphism; also, $fm = gm$. We shall use the description of $\mathrm{Equ}(f,g)$ given by I.6.2 and hence try to prove that $\mathrm{Im}(m \times m) = g^{-1}f \wedge \varepsilon$. Since $\mathrm{Im}\, m \leq \mathrm{Equ}(f,g)$, we already know that $\mathrm{Im}(m \times m) \leq g^{-1}f \wedge \varepsilon$.

For each $i \in I$, we have, by 2.6 and (C_3') :

$$\tilde{x}_i^s (g^{-1}f \wedge \varepsilon) = \tilde{x}_i^s (g^{-1}f) \wedge \tilde{x}_i^s \varepsilon = x_i^{-1}g^{-1} f x_i \wedge x_i^{-1}x_i =$$
$$= g_i^{-1}y_i^{-1}y_i f_i \wedge x_i^{-1}x_i =$$
$$= (\bigvee_{j \geq i} g_i^{-1}y_{ij}^{-1}y_{ij}f_i) \wedge (\bigvee_{k \geq i} x_{ik}^{-1}x_{ik}) =$$
$$= \bigvee_{j,k \geq i} (g_i^{-1}y_{ij}^{-1}y_{ij}f_i \wedge x_{ik}^{-1}x_{ik}) =$$
$$= \bigvee_{t \geq i} (g_i^{-1}y_{it}^{-1}y_{it}f_i \wedge x_{it}^{-1}x_{it}) \quad \text{[by directedness]} =$$
$$= \bigvee_{t \geq i} (x_{it}^{-1}g_t^{-1}f_t x_{it} \wedge x_{it}^{-1}x_{it}) =$$

$$= \bigvee_{t\geq i} \tilde{x}_{it}^{s}(g_t^{-1}f_t \wedge \epsilon) = \bigvee_{t\geq i} \tilde{x}_{it}^{s}\, \mathrm{Im}(m_t \times m_t)$$

since $m_t \in \mathrm{Equ}(f_t, g_t)$. Therefore

$$\begin{aligned}(g^{-1}f \wedge \epsilon)\wedge \mathrm{Im}\, \tilde{x}_i &= (\tilde{x}_i)_s\, \tilde{x}_i^{s}(g^{-1}f \wedge \epsilon) = \bigvee_{t\geq i} (\tilde{x}_i)_s\, \tilde{x}_{it}^{s}\, \mathrm{Im}(m_t \times m_t) = \\ &= \bigvee_{t\geq i} (\tilde{x}_t)_s(\tilde{x}_{it})_s(\tilde{x}_{it})^{s}\, \mathrm{Im}(m_t \times m_t) \leq \\ &\leq \bigvee_{t\geq i} (\tilde{x}_t)_s\, \mathrm{Im}(m_t \times m_t) = \bigvee_{t\geq i} \mathrm{Im}(x_t m_t \times x_t m_t) \leq \\ &\leq \mathrm{Im}(m \times m)\end{aligned}$$

since $\mathrm{Im}\, x_t m_t = \mathrm{Im}\, me_t \leq \mathrm{Im}\, m$. Then it follows from (C_3') and 2.7 that $g^{-1}f \wedge \epsilon = \bigvee_{i\in I} ((g^{-1}f \wedge \epsilon)\wedge \mathrm{Im}\, \tilde{x}_i) \leq \mathrm{Im}(m \times m)$, q.e.d.

3. We now turn to the preservation of finite products. First we claim that it suffices to prove that the functor $-\pi A : \mathcal{G} \longrightarrow \mathcal{G}$ preserves directed colimits, for every $A \in \mathcal{G}$. This will indeed yield a natural isomorphism $\varinjlim(\mathfrak{X}\, \pi A) \cong (\varinjlim \mathfrak{X})\, \pi A$ for every direct system $\mathfrak{X}$, hence, for any two $\mathfrak{X}, \mathfrak{Y} \in [I, \mathcal{G}]$, natural isomorphisms

$$\varinjlim_{(i,j)\in I\pi I} (X_i\, \pi\, Y_j) \cong \varinjlim_{i\in I} (X_i\, \pi\, \varinjlim \mathfrak{Y}) \cong \varinjlim \mathfrak{X}\, \pi\, \varinjlim \mathfrak{Y} ;$$

and since I is directed, the diagonal is cofinal in $I \pi I$, so that there is a natural isomorphism

$$\varinjlim(\mathfrak{X}\, \pi\, \mathfrak{Y}) = \varinjlim_{i\in I} (X_i\, \pi\, Y_i) \cong \varinjlim_{(i,j)\in I\pi I} (X_i\, \pi\, Y_j) .$$

4. Now our functor $-\pi A$ preserves pullbacks (as readily seen) and regular decompositions (by I.1.12). Also, by I.1.13, I.3.10, for any f, $\mathrm{Im}(f\, \pi\, 1_A) = p^{s}\, \mathrm{Im}\, f$, where p is a projection, so that our functor also preserves directed unions of subobjects, by (C_3') .

Then let $\mathfrak{X}$ be a direct system (over I); let $\mathfrak{Y} = \mathfrak{X}\, \pi A$, so that $Y_i = X_i\, \pi A$ etc. It is clear that $(x_i\, \pi\, 1_A)_{i\in I}$ is a cocompatible family for $\mathfrak{Y}$; hence there is a morphism $t : Y \longrightarrow X\, \pi A$ such that $x_i\, \pi\, 1_A = t y_i$ for all i . Clearly t is natural in $\mathfrak{X}$; we want to show that it is an isomorphism.

Since our functor preserves directed unions of subobjects, $\bigvee_{i\in I} \operatorname{Im} x_i = 1$ implies $\bigvee_{i\in I} \operatorname{Im}(x_i \pi 1_A) = \operatorname{Im}(1 \pi 1_A) = 1$. Hence

$$\operatorname{Im} t = t_s(\bigvee_{i\in I} \operatorname{Im} y_i) = \bigvee_{i\in I} \operatorname{Im} ty_i = \bigvee_{i\in I} \operatorname{Im}(x_i \pi 1_A) = 1$$

and t is a regular epimorphism.

On the other hand, our functor preserves pullbacks, hence also congruences, as well as directed unions of subobjects, and therefore $\ker x_i = \bigvee_{j\geq i} \ker x_{ij}$ (2.6) implies $\ker(x_i \pi 1_A) = \bigvee_{j\geq i} \ker(x_{ij} \pi 1_A)$. Hence

$$\begin{aligned}
\ker t &= \ker t \wedge (\bigvee_{i\in I} \operatorname{Im} \tilde{y}_i) = \bigvee_{i\in I} (\ker t \wedge \operatorname{Im} \tilde{y}_i) = \\
&= \bigvee_{i\in I} (\tilde{y}_i)_s (\tilde{y}_i)^s \ker t = \bigvee_{i\in I} (\tilde{y}_i)_s \ker ty_i = \\
&= \bigvee_{i\in I} (\tilde{y}_i)_s \ker(x_i \pi 1_A) = \\
&= \bigvee_{i\in I} (\tilde{y}_i)_s (\bigvee_{j\geq i} \ker(x_{ij} \pi 1_A)) = \bigvee_{i\in I} (\tilde{y}_i)_s (\bigvee_{j\geq i} \ker y_{ij}) = \\
&= \bigvee_{i\in I} (\tilde{y}_i)_s \ker y_i \leq \varepsilon .
\end{aligned}$$

Thus t is also a monomorphism. Therefore it is an isomorphism, q.e.d.

The proof of the theorem is now complete.

4. ADDITIONAL PROPERTIES OF DIRECTED COLIMITS.

1. We now let $\mathcal{G}$ be a C_3 regular category. From the property 2.6 that for any direct system $X : I \longrightarrow \mathcal{G}$, $\ker x_i = \bigvee_{j\geq i} \ker x_{ij}$ for all i , it is easy to derive a number of additional properties.

First we have a very 'set-like' result, which complement the construction in the proof of 2.6 and could also be used in the last part of the proof above.

Proposition 4.1. Let $\mathfrak{X}$ be a direct system over I in a C_3 regular category, and $(a_i)_{i\in I}$ a cocompatible family for $\mathfrak{X}$ inducing at the colimit a morphism a. Then:

i) a is a regular epimorphism if and only if $\bigvee_{i\in I} \operatorname{Im} a_i = 1$;

ii) a is a monomorphism if and only if $\ker a_i = \bigvee_{j\geq i} \ker x_{ij}$ for for every i;

iii) a is an isomorphism if and only if both conditions hold.

Proof. First $\operatorname{Im} a = \bigvee_{i\in I} \operatorname{Im} a_i$ by I.6.6, which proves i). In view of 2.6, ii) says that a is a monomorphism if and only if $\ker a_i = \ker x_i$ for every i. Since $a_i = ax_i$ this is certainly necessary. If conversely $\ker a_i = \ker x_i$ for every i, then the familiar argument

$$\ker a = \bigvee_{i\in I}(\ker a \wedge \operatorname{Im} \tilde{x}_i) = \bigvee_{i\in I} (\tilde{x}_i)_s \ker ax_i =$$
$$= \bigvee_{i\in I} (\tilde{x}_i)_s \ker x_i \leq \epsilon$$

shows that a is a monomorphism. Finally, iii) follows from i) and ii).

Then we have two equalizer properties.

Proposition 4.2. In a C_3 regular category, Gray's condition $\mathfrak{F}_2$ holds; in other words, for every direct system $\mathfrak{X}$, $x_i f = x_i g$ implies $\bigvee_{j\geq i} \operatorname{Equ}(x_{ij}f, x_{ij}g) = 1$.

Proof. It follows from I.6.2 that

$$\bigvee_{j\geq i} \operatorname{Equ}(x_{ij}f, x_{ij}g) = \bigvee_{j\geq i} \Delta^s (g^{-1}x_{ij}^{-1}x_{ij}f) = \Delta^s(g^{-1}(\bigvee_{j\geq i} x_{ij}^{-1}x_{ij})f) =$$
$$= \Delta^s(g^{-1}x_i^{-1}x_i f) = \operatorname{Equ}(x_i f, x_i g) = 1 .$$

Proposition 4.3. Let $\mathfrak{X}$ be a direct system over I in a C_3 regular category, and $f,g : \lim \mathfrak{X} \longrightarrow A$. Then $\operatorname{Equ}(f,g) = \bigvee_{i\in I}(x_i)_s \operatorname{Equ}(fx_i, gx_i)$.

Proof. Take $m \in \mathrm{Equ}(f,g)$, $n \in \mathrm{Equ}(fx_i, gx_i)$. Since $fx_i n = gx_i n$, there is a commutative square $x_i n = mt$; we claim it is in fact a pullback. Assume that $x_i a = mb$. Then $fx_i a = fmb = gmb = gx_i a$, so that $a = nu$ for some u; also, $mtu = x_i nu = x_i a = mb$, and $b = tu$. This factorization is unique since n is a monomorphism.

Then it follows that $(x_i)^s \mathrm{Equ}(f,g) = \mathrm{Equ}(fx_i, gx_i)$. Hence $(x_i)_s\ \mathrm{Equ}(fx_i, gx_i) = (x_i)_s (x_i)^s\ \mathrm{Equ}(f,g) = \mathrm{Equ}(f,g) \wedge \mathrm{Im}\ x_i$. The result follows since $\bigvee_{i \in I} \mathrm{Im}\ x_i = 1$.

2. Finally we show that additional conditions insure additional good behavior of directed colimits, with regards to [not necessarily finite] products. These are, first, Gray's condition $\mathfrak{Z}_1$:

$\mathfrak{Z}_1$: if $((\underline{x}_i)_{i \in I_\lambda})_{\lambda \in \Lambda}$ is a non-empty family of [non-empty] directed families of subobjects of the same $A \in \mathcal{C}$, then

$$\bigwedge_{\lambda \in \Lambda} \Big(\bigvee_{i \in I_\lambda} \underline{x}_i \Big) = \bigvee_{\tau \in T} \Big(\bigwedge_{\lambda \in \Lambda} \underline{x}_{\tau\lambda} \Big)$$

where $T = \prod_{\lambda \in \Lambda} I_\lambda$.

This condition is formulable in any category with intersections. We shall always assume in the above that the sets I_λ are pairwise disjoint, and write $\tau\lambda$ instead of τ_λ (to avoid seventh order subscripts). A complete C_3 regular category satisfying $\mathfrak{Z}_1$ is called a C_4 regular category. Examples include of course C_4 abelian categories, and finitary varieties, in which $\mathfrak{Z}_1$ becomes the familiar $\cap$-$\cup$ distributivity (in its general form). We note that $\mathfrak{Z}_1$ implies A.B.5 but not (C_3'); yet the axioms of C_4 regular categories become redundant in yet another way, since by I.6.9 cocompleteness can be replaced by the existence of coproducts under either (L) or minor size restrictions (see also the results in [2] for other implications, under stronger size restrictions).

The other condition is that $\mathcal{C}$ be C_1^*, i.e. any product of regular epimorphisms is a regular epimorphism. The finite version of that condition would evaporate, by I.1.12. The condition itself holds in any variety.

Using these conditions, we have:

Theorem 4.4. Let $\mathcal{C}$ be a C_4 regular category and $(\mathcal{X}^\lambda)_{\lambda\in\Lambda}$ be a non-empty family of direct systems $\mathcal{X}^\lambda : I_\lambda \longrightarrow \mathcal{C}$. The morphisms $x'_\tau = \prod_{\lambda\in\Lambda} x_{\tau\lambda}$, $\tau \in T = \prod_{\lambda\in\Lambda} I_\lambda$, induce a natural monomorphism:

$$t : \varinjlim_{\tau\in T} \prod_{\lambda\in\Lambda} X^\lambda_{\tau\lambda} \longrightarrow \prod_{\lambda\in\Lambda} \varinjlim \mathcal{X}^\lambda$$

which is in fact an isomorphism if all $\mathcal{X}^\lambda$ are monic, or if $\mathcal{C}$ is C_1^*.

Of course we assume that the I_λ are pairwise disjoint, which allows us to write X_i etc. instead of X_i^λ $(i \in I_\lambda)$. Also note that under the coordinatewise preorder T is a directed preordered set; a direct system $\mathcal{X} : T \longrightarrow \mathcal{C}$ is defined by $X_\tau = \prod_{\lambda\in\Lambda} X_{\tau\lambda}$, $x_{\sigma\tau} = \prod_{\lambda\in\Lambda} x_{\sigma\lambda,\tau\lambda}$ $(\sigma \leq \tau)$, giving the new [directed] colimit that appears in the theorem.

3. We begin the proof with the following generalization of 2.7:

Lemma 4.5. Let $(A_\lambda)_{\lambda\in\Lambda}$ be a non-empty family of objects of $\mathcal{C}$ and, for each λ, $(f_i)_{i\in I_\lambda}$ be a family of morphisms of codomain A such that $(\operatorname{Im} f_i)_{i\in I_\lambda}$ is directed with $\bigvee_{i\in I_\lambda} \operatorname{Im} f_i = 1$. For each $\tau\in T$, let $f_\tau = \prod_{\lambda\in\Lambda} f_{\tau\lambda}$. If all f_i are monomorphisms, or if $\mathcal{C}$ is C_1^*, then $\bigvee_{\tau\in T} \operatorname{Im} f_\tau = 1$.

Proof. We consider first the case when all f_i are monomorphisms. For each $i \in I_\mu$, $\tau\in T$, put $f_i : Y_i \longrightarrow A_\mu$ and consider the diagram:

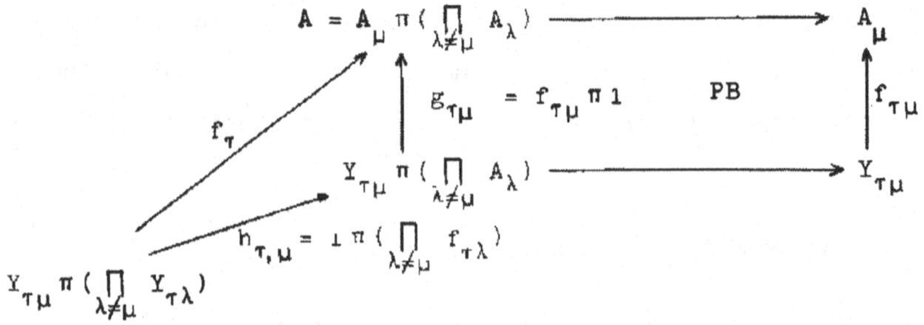

where the horizontal maps are projections. The diagram commutes, in fact the square is a pullback (I.1.13), and the new maps $g_{\tau\mu}$, $h_{\tau,\mu}$ are monomorphisms. We note that the square still serves if $\tau\mu$ is replaced by $i \in I_\mu$ (and then we denote $f_i \pi 1$ by g_i , instead of $g_{\tau\mu}$).

We see on the diagram that f_τ factors through all $g_{\tau\mu}$ $(\mu \in \Lambda)$. In fact it is an intersection of that family. Indeed let $u : Z \longrightarrow A$ factor through every $g_{\tau\mu}$ (with τ given); write $u = \bigtimes_{\lambda\in\Lambda} u_\lambda$, $u_\lambda : Z \longrightarrow A$. For each μ, $u = g_{\tau\mu} v_\mu$ for some $v_\mu = w_\mu \times (\bigtimes_{\lambda\neq\mu} w_{\mu\lambda})$; note that $w_\mu : Z \longrightarrow Y_{\tau\mu}$, $w_{\mu\lambda} : Z \longrightarrow A_\lambda$ if $\lambda \neq \mu$. Since $u = g_{\tau\mu} v_\mu$ we see that $u_\mu = f_{\tau\mu} w_\mu$, $u_\lambda = w_{\mu\lambda}$ if $\lambda \neq \mu$. Hence $v_\mu = w_\mu \times (\bigtimes_{\lambda\neq\mu} u_\lambda) = w_\mu \times (\bigtimes_{\lambda\neq\mu} f_{\tau\lambda} w_\lambda) = h_{\tau,\mu} \cdot (\bigtimes_{\lambda\in\Lambda} w_\lambda) = h_{\tau,\mu} w$, say . Therefore $u = g_{\tau\mu} v_\mu = g \quad h_{\tau,\mu} w = f_\tau w$, i.e. u factors through f_τ . Thus we do have an intersection and it follows that
$\operatorname{Im} f_\tau = \bigwedge_{\lambda\in\Lambda} \operatorname{Im} g_{\tau\lambda}$.

On the other hand, let $p_\mu : A \longrightarrow A_\mu$ be the projection. For each $i \in I_\mu$ the pullback above (with $\tau\mu$ replaced by i) yields $\operatorname{Im} g_i = p_\mu^s \operatorname{Im} f_i$; by (C_3^*) and the hypothesis, it follows that $\bigvee_{i\in I_\mu} \operatorname{Im} g_i = 1$ for each $\mu \in \Lambda$. Then, by $\mathfrak{I}_1$:

$$\bigvee_{\tau\in T} \operatorname{Im} f_\tau = \bigvee_{\tau\in T} (\bigwedge_{\lambda\in\Lambda} \operatorname{Im} g_{\tau\lambda}) = \bigwedge_{\lambda\in\Lambda} (\bigvee_{i\in I_\lambda} \operatorname{Im} g_i) = 1 .$$

This takes care of the case when all f_i are monomorphisms. In the general case, we also assume that G is C_1^*, so that products preserve regular decompositions. Thus we can select for each i a regular decomposition (m_i, p_i) of f_i and obtain a regular decomposition (m_τ, p_τ) of f_τ, with $m_\tau = \prod_{\lambda\in\Lambda} m_{\tau\lambda}$. Since $\operatorname{Im} m_i = \operatorname{Im} f_i$ for every i, it follows from the above that $\bigvee_{\tau\in T} \operatorname{Im} m_\tau = 1$ and the result again holds since $\operatorname{Im} f_\tau = \operatorname{Im} m_\tau$.

4. We now prove the theorem. We want to show that the morphism t induced by all $x'_\tau = \prod_{\lambda\in\Lambda} x_{\tau\lambda}$ is a monomorphism and in some cases an isomorphism. We use 4.1. If all x_i are monomorphisms (i.e. if all $\mathfrak{X}^\lambda$ are monic), or if G is C_1^*, then the lemma applies to $((x_i)_{i\in I_\lambda})_{\lambda\in\Lambda}$ and yields $\bigvee_{\tau\in T} \operatorname{Im} x'_\tau$, so that t is a regular epimorphism.

We now show that t is a monomorphism, without using C_1^*. First note that the result is trivial when all $\mathfrak{X}^\lambda$ are monic, for then all x'_τ are monomorphisms; in that case, the proof of the theorem is over.

In the general case, take $\tau \in T$. For each λ and each $i \in I_\lambda$ with $i \geq \tau\lambda$, select monomorphisms $m_{\tau\lambda} : K_{\tau\lambda} \longrightarrow X_{\tau\lambda} \pi X_{\tau\lambda}$, $m_{\tau\lambda,i} : K_{\tau\lambda,i} \longrightarrow X_{\tau\lambda} \pi X_{\tau\lambda}$ such that $\ker x_{\tau\lambda} = \operatorname{Im} m_{\tau\lambda}$, $\ker x_{\tau\lambda,i} = \operatorname{Im} m_{\tau\lambda,i}$. Since $(\operatorname{Im} m_{\tau\lambda,i})_{i\in I}$ is a directed family of subobjects, there is a monic direct system $\mathcal{K}^\lambda$ over $\{i \in I_\lambda \,;\, i \geq \tau\lambda\}$ with objects $K_{\tau\lambda,i}$. Since $\operatorname{Im} m_{\tau\lambda,i} \leq \operatorname{Im} m_{\tau\lambda}$, we have a clearly cocompatible family of monomorphisms $K_{\tau\lambda,i} \longrightarrow K_{\tau\lambda}$; since in fact $\operatorname{Im} m_{\tau\lambda} = \bigvee_{i\geq\tau\lambda} \operatorname{Im} m_{\tau\lambda,i}$, the induced monomorphism $\varinjlim \mathcal{K}^\lambda \longrightarrow K_{\tau\lambda}$ is an isomorphism, by 4.1.

Since we have already proved the theorem in the case of monic direct systems, we can apply it to the family $(\mathcal{K}^\lambda)_{\lambda\in\Lambda}$; we obtain an isomorphism

$$\varinjlim_{\sigma\in\Sigma} \prod_{\lambda\in\Lambda} K_{\tau\lambda,\sigma\lambda} \cong \prod_{\lambda\in\Lambda} K_{\tau\lambda},$$

where $\Sigma = \prod_{\lambda\in\Lambda} \{i \in I_\lambda ; i \geq \tau\lambda\} = \{\sigma \in T ; \sigma \geq \tau\}$, induced by all $\prod_{\lambda\in\Lambda}$ of monomorphisms $K_{\tau\lambda,\sigma\lambda} \longrightarrow K_{\tau\lambda}$. It follows that the morphism induced to the colimit by the cocompatible family $(\prod_{\lambda\in\Lambda} m_{\tau\lambda,\sigma\lambda})_{\sigma\in\Sigma}$ is equivalent (as a monomorphism) to $\prod_{\lambda\in\Lambda} m_{\tau\lambda}$; then, by I.6.6,

$$\bigvee_{\sigma\geq\tau} (\operatorname{Im} \prod_{\lambda\in\Lambda} m_{\tau\lambda,\sigma\lambda}) = \operatorname{Im} \prod_{\lambda\in\Lambda} m_{\tau\lambda} .$$

We now remember that $x_{\tau\sigma} = \prod_{\lambda\in\Lambda} x_{\tau\lambda,\sigma\lambda}$, $x'_\tau = \prod_{\lambda\in\Lambda} x_{\tau\lambda}$; since products preserve kernels, hence also congruences, we have in fact proved that $\ker x'_\tau = \bigvee_{\sigma\geq\tau} \ker x_{\tau\sigma}$. Then it follows from 4.1 that t is a monomorphism, and this completes the proof of the theorem.

III. SHEAVES IN REGULAR CATEGORIES

In this part we study categories $\mathfrak{F}(X,\mathfrak{G})$ of sheaves in a regular category $\mathfrak{G}$ (which we assume is at least C_4) over an arbitrary topological space or Grothendieck topology X.

Our first result is the existence of the associated sheaf of a presheaf, i.e. $\mathfrak{F}(X,\mathfrak{G})$ is coreflective in the category $\mathfrak{P}(X,\mathfrak{G})$ of presheaves. The previous results of that kind are due to Gray [10],[11] (see also [31]), and, in the exact case, to Heller and Rowe [16], extending an older result of Grothendieck [15] . Gray's proof of existence is similar to that of the adjoint functor theorem (except that solution sets are neatly bypassed). Heller and Rowe's proof is constructive: one basic construction is iterated to build increasingly sheaf-like presheaves, which eventually terminates at the associated sheaf. In either case, some restriction is made on the size of the category one works in, such as being locally small or having a set of generators. We prove that, when $\mathfrak{G}$ is a C_4 regular category, and X is any Grothendieck topology, Heller and Rowe's construction terminates after two steps; this answers a conjecture made by Gray [12] and in particular proves the existence of the associated sheaf. The assumption on $\mathfrak{G}$ has more cocompleteness than in the other results, but involves no size restriction, and the result is new even in the case of a C_4 abelian category.

Under the same hypothesis, we show that $\mathfrak{F}(X,\mathfrak{G})$ is a C_3 regular category. If furthermore $\mathfrak{G}$ is C_1^* and X is a topological space, then the stalk functor reflects isomorphisms [which means that it is cotripleable, as VanOsdol pointed out to us] so that all finite limits, regular decompositions and colimits in $\mathfrak{F}(X,\mathfrak{G})$ can safely be computed on the stalks.

Related and additional results will be found in VanOsdol's contribution to this volume. We owe much to VanOsdol, for discussions, and e.g. suggesting that 3.1 below might hold and bring an answer to Gray's conjecture. In addition, the results in section 4 were first proved by him in the case of varieties [35].

Our exposition follows that of [14], except for the inclusion of the details of Heller and Rowe's construction and the rather straightforward extension to Grothendieck topologies at the beginning [independently suggested by VanOsdol and Heller].

1. GROTHENDIECK TOPOLOGIES AND SHEAVES.

In this section we recall the basic definitions concerning Grothendieck topologies and sheaves thereon, and set forth some notation.

1. A Grothendieck topology X is a small category $\mathfrak{U}(X)$ together with a set $\mathfrak{C}$ of coterminal families of morphisms of $\mathfrak{U}(X)$ ('coverings') satisfying the following conditions (in which $\mathfrak{C}(U)$ denotes the set of all coverings of [codomain] U):

i) $\{\alpha\} \in \mathfrak{C}$ for every isomorphism $\alpha \in \mathfrak{U}(X)$;

ii) if $(\alpha_i)_{i\in I} \in \mathfrak{C}(U)$, $\alpha_i : U_i \longrightarrow U$, and for every $i \in I$ $(\beta_{ij})_{j\in J_i} \in \mathfrak{C}(U_i)$, then $(\alpha_i\beta_{ij})_{j\in J_i,\, i\in I} \in \mathfrak{C}(U)$;

iii) if $(\alpha_i)_{i\in I} \in \mathfrak{C}(U)$ and $\beta : V \longrightarrow U \in \mathfrak{U}(X)$, then for each i there exists a pullback $\beta\alpha_i' = \alpha_i\gamma_i$, and $(\alpha_i')_{i\in I} \in \mathfrak{C}(V)$.

The prime example of a Grothendieck topology is given by any topology in the usual sense, i.e. the family $\mathfrak{U}(X)$ of all open subsets of a topological space X. Then $\mathfrak{U}(X)$ is made into a category in the obvious way (the morphisms being all inclusion maps between objects, i.e. elements, of $\mathfrak{A}(X)$), and $\mathfrak{C}(U)$ is the set of all families of inclusion maps $U_i \longrightarrow U$ $(i \in I)$ such that $U_i \in \mathfrak{A}(X)$ and $\bigcup_{i \in I} U_i = U$. Any small regular category $\mathcal{G}$ provides another example (which will not be used here): let $\mathfrak{A} = \mathcal{G}$ and $\mathfrak{C}$ be the set of all families that consist of just one regular epimorphism. If (C_3') holds in $\mathcal{G}$, another $\mathfrak{C}$ is defined by: $(\alpha_i)_{i \in I} \in \mathfrak{C}$ if and only if $(\mathrm{Im}\, \alpha_i)_{i \in I}$ is directed and $\bigvee_{i \in I} \mathrm{Im}\, \alpha_i = 1$. More examples can be found e.g. in [6].

2. If X is any Grothendieck topology, then, for each $U \in \mathfrak{A}(X)$, $\mathfrak{C}(U)$ can be made into a directed preordered set as follows. If $\mathcal{C} = (\alpha_i)_{i \in I}$, $\alpha_i : U_i \longrightarrow U$ and $\mathcal{D} = (\beta_j)_{j \in J}$, $\beta_j : V_j \longrightarrow U$ are in $\mathfrak{C}(U)$, say that $\mathcal{D}$ <u>refines</u> $\mathcal{C}$, and write $\mathcal{C} \leq \mathcal{D}$, in case there exists a mapping $\psi : J \longrightarrow I$ and morphisms $\psi_j : V_j \longrightarrow U_{\psi j}$ such that $\beta_j = \alpha_{\psi j}\, \psi_j$ for every $j \in J$. If $\mathcal{D} \leq \mathcal{E} = (\gamma_k)_{k \in K}$, with $\chi : K \longrightarrow J$ and morphisms χ_k serving in the definition (i.e., $\gamma_k = \beta_{\chi k}\, \chi_k$ for all k), then $\omega = \psi\chi : K \longrightarrow I$ and $\omega_k = \psi_{\chi k}\chi_k$ are such that $\gamma_k = \beta_{\chi k}\, \chi_k = \alpha_{\psi\chi k}\, \psi_{\chi k}\, \chi_k = \alpha_{\omega k}\, \omega_k$, and therefore $\mathcal{C} \leq \mathcal{E}$, which shows that $\leq$ is transitive; it is clearly reflexive. (In fact, we have implicitly defined morphisms in $\mathfrak{C}(U)$ and made it into a category).

To show directedness, start with $\mathcal{C}$ and $\mathcal{D}$ as above (except that $\mathcal{D}$ need not refine $\mathcal{C}$). By iii) there is for each i and j a pullback $\alpha_i\, \beta'_{ij} = \beta_j\, \alpha'_{ij}$. The family $\gamma_{ij} : U_i * V_j \longrightarrow U$ $((i,j) \in I \pi J)$ defined by $\gamma_{ij} = \beta_j \alpha'_{ij}$ is a covering of U by ii), since by iii) $(\alpha'_{ij})_{i \in I}$ is in $\mathfrak{C}(V_j)$ for every j; we denote it by $\mathcal{C} * \mathcal{D}$. [Other notations for $U_i * V_j$ are $U_i \times_U V_j$ and $U_i \cap V_j$. The notation $\mathcal{C} * \mathcal{D}$ is legitimate since in the small category $\mathfrak{A}(X)$ we can once and for all make a selection of pullbacks which covers all existence cases postulated by iii).] To see that $\mathcal{C} \leq \mathcal{C} * \mathcal{D}$ it suffices to consider the projection $\psi : I \pi J \longrightarrow I$ and define $\psi_{ij} = \beta'_{ij} : U_i * V_j \longrightarrow U_i$. Similarly, $\mathcal{D} \leq \mathcal{C} * \mathcal{D}$. Thus, for each $U \in \mathfrak{A}(X)$ we now have a [non-empty, by i)] directed preordered set $\mathfrak{C}(U)$.

If $C = (\alpha_i)_{i\in I} \in \mathfrak{C}(U)$ and $\beta : V \longrightarrow U$, the covering postulated by iii) will be denoted by $C * V$ (when there can be no confusion on β). We note that in the above $C * \mathfrak{D}$ is obtained by composing (as in ii)) $\mathfrak{D}$ and all $C * V_j$ (and also by composing C and all $\mathfrak{D} * U_i$).

3. Let X be a Grothendieck topology and $\mathfrak{a}$ be any category which has products. A <u>presheaf</u> on X with values in $\mathfrak{a}$ is a contravariant functor $P : \mathfrak{U}(X) \longrightarrow \mathfrak{a}$; these form a category $\mathcal{P}(X,\mathfrak{a})$.

For each $P \in \mathcal{P}(X,\mathfrak{a})$ and $C \in \mathfrak{C}(U)$ we have a canonical diagram

$$P(U) \xrightarrow{u} P(C) \overset{f}{\underset{g}{\rightrightarrows}} P(C*C)$$

defined as follows. Put $C = (\alpha_i)_{i\in I}$, $\alpha_i : U_i \longrightarrow U$ and let $\alpha_j \xi_{jk} = \alpha_k \eta_{jk}$ be the pullbacks defining $C * C$. Then $P(C) = \prod_{i\in I} P(U_i)$ and $P(C*C) = \prod_{j,k\in I} P(U_j * U_k)$. The morphisms u,f,g are induced by the obvious 'restriction maps', namely: $u = u_C^P = \underset{i\in I}{\times} P(\alpha_i)$, $f = f_C^P = \prod_{j\in I}(\underset{k\in I}{\times} P(\xi_{jk}))$, $g = g_C^P = \prod_{k\in I}(\underset{j\in I}{\times} P(\eta_{jk}))$; if we use the letter π to denote any projection from a product, as we shall do from here on, e.g. $\pi_i : P(C) \longrightarrow P(U_i)$, then we see that $\pi_i u = P(\alpha_i)$, $\pi_{jk} f = P(\xi_{jk})\pi_j$, $\pi_{jk} g = P(\eta_{jk})\pi_k$. The reader should verify that $fu = gu$.

The presheaf P is called a <u>monopresheaf</u> if u is a monomorphism for all C, and a <u>sheaf</u> if $u \in \mathrm{Equ}(f,g)$ for all C. The category of all sheaves on X with values in $\mathfrak{a}$ (a full subcategory of $\mathcal{P}(X,\mathfrak{a})$) will be denoted by $\mathfrak{F}(X,\mathfrak{a})$. It is defined in terms of products and equalizers, and since these commute with limits and limits in the functor category $\mathcal{P}(X,\mathfrak{a})$ are eveluated pointwise, it follows that $\mathfrak{F}(X,\mathfrak{a})$ is a complete subcategory of $\mathcal{P}(X,\mathfrak{a})$ (i.e. admits all existing limits).

2. THE HELLER AND ROWE CONSTRUCTION OF THE ASSOCIATED SHEAF.

(1.) Let X be any Grothendieck topology and $\mathfrak{G}$ be a complete category. Then we know that $\mathfrak{F}(X,\mathfrak{G})$ is a complete subcategory of $\mathcal{P}(X,\mathfrak{G})$ and one may feel that it will take very little for $\mathfrak{F}(X,\mathfrak{G})$ to be coreflexive in $\mathcal{P}(X,\mathfrak{G})$. A look at the existing results of that sort shows that this first impression may be misleading. It takes a complete well-powered category $\mathfrak{G}$ having directed colimits and satisfying Gray's condition $\mathfrak{F}_1,\mathfrak{F}_2$, for the existence of associated sheaves to be established by a reasonably short argument, similar to the proof of the adjoint functor theorem [10], [31]. A more explicit construction was given, when $\mathfrak{G}$ is a complete exact category having a projective generator and directed colimits which are exact, by Heller and Rowe [16]; in this construction, a presheaf P' is explicitely constructed from any given presheaf P, and when the construction is repeated sufficiently many times (by ordinal induction) it eventually terminates at the associated sheaf.

It was conjectured by Gray in [12] that in most good categories Heller and Rowe's construction should yield the associated sheaf in two steps. We shall prove this is indeed the case when $\mathfrak{G}$ is a regular C_4 category. First, we recall Heller and Rowe's construction; in what follows, $\mathfrak{G}$ is a complete category having directed colimits and X is any Grothendieck topology; $P \in \mathcal{P}(X,\mathfrak{G})$ is given.

(2.) For each $C \in \mathfrak{C}(U)$, we have a canonical diagram

$$P(U) \xrightarrow{u} P(C) \underset{g}{\overset{f}{\rightrightarrows}} P(C * C) ;$$

let $u_C^* : E_C(U) \longrightarrow P(C) \in \mathrm{Equ}_{\mathfrak{G}}(f,g)$. Since $fu = fu$, we have $u_C = u_C^*\, c_C(U)$ for some unique $c_C(U) : P(U) \longrightarrow E_C(U)$.

We now organize the objects $E_C(U)$ into a direct system over

$\mathfrak{C}(U)$. Let $\mathcal{C}, \mathcal{D} \in \mathfrak{C}(U)$ satisfy $\mathcal{C} \leq \mathcal{D}$. Put $\mathcal{C} = (\alpha_i)_{i \in I}$, $\mathcal{D} = (\beta_j)_{j \in J}$; then there exist a mapping $\psi : J \longrightarrow I$ and morphisms ψ_p $(p \in J)$ such that $\beta_p = \alpha_{\psi p} \psi_p$ for all p . Also, let $\alpha_j \xi_{jk} = \alpha_k \eta_{jk}$, $\beta_p \xi_{pq} = \beta_q \eta_{pq}$ $(j, k \in I$, $p, q \in J)$ be pullbacks; for each $p, q \in J$, $\alpha_{\psi p} \psi_p \xi_{pq} = \alpha_{\psi q} \psi_q \eta_{pq}$, so that there exists a unique morphism ψ_{pq} such that $\psi_p \xi_{pq} = \xi_{\psi p, \psi q} \psi_{pq}$, $\psi_p \eta_{pq} = \eta_{\psi p, \psi q} \psi_{pq}$ (that is, $\mathcal{C} * \mathcal{C} \leq$ $\leq \mathcal{D} * \mathcal{D}$). We can then define morphisms $P'(\psi) : P(\mathcal{C}) \longrightarrow P(\mathcal{D})$, $P''(\psi) : P(\mathcal{C} * \mathcal{C}) \longrightarrow P(\mathcal{D} * \mathcal{D})$ by:

$$P'(\psi) = \prod_{i \in I} \Big(\bigtimes_{p \in \psi^{-1} i} P(\psi_p) \Big)$$

$$P''(\psi) = \prod_{\substack{j \in I \\ k \in I}} \Big(\bigtimes_{\substack{p \in \psi^{-1} j \\ q \in \psi^{-1} k}} P(\psi_{pq}) \Big)$$

(so that $\pi_p P'(\psi) = P(\psi_p) \pi_{\psi p}$, $\pi_{pq} P''(\psi) = P(\psi_{pq}) \pi_{\psi p, \psi q}$).

<u>Lemma</u> 2.1. $f_{\mathcal{D}} P'(\psi) = P''(\psi) f_{\mathcal{C}}$, $g_{\mathcal{D}} P'(\psi) = P''(\psi) g_{\mathcal{C}}$ and furthermore $P'(\psi) u^*_{\mathcal{C}}$ depends only on $\mathcal{C}$ and $\mathcal{D}$ and not on the choice of ψ .

<u>Proof</u>. For each $p, q \in J$, composing $f_{\mathcal{D}} P'(\psi)$ and $P''(\psi) f_{\mathcal{C}}$ with π_{pq} yields, respectively, $P(\xi_{pq}) P(\psi_p) \pi_{\psi p}$ and $P(\psi_{pq}) P(\xi_{\psi p, \psi q}) \pi_{\psi p}$; due to the relation $\psi_p \xi_{pq} = \xi_{\psi p, \psi q} \psi_{pq}$ above, these are always equal, which proves the first formula. The second one is proved similarly. For the last part, let $\psi' : J \longrightarrow I$, ψ'_p $(p \in J)$ be another mapping and family of morphisms such that $\beta_p = \alpha_{\psi' p} \psi'_p$ for all p .

Since $\alpha_{\psi p} \psi_p = \alpha_{\psi' p} \psi'_p$, we have $\psi_p = \xi_{\psi p, \psi' p} \psi''_p$, $\psi'_p = \eta_{\psi p, \psi' p} \psi''_p$ for some ψ''_p . Hence for each $p \in J$

$$\pi_p P'(\psi) u^*_{\mathcal{C}} = P(\psi_p) \pi_{\psi p} u^*_{\mathcal{C}} = P(\psi''_p) P(\xi_{\psi p, \psi' p}) \pi_{\psi p} u^*_{\mathcal{C}} =$$

$$= P(\psi''_p) \pi_{\psi p, \psi' p} f_{\mathcal{C}} u^*_{\mathcal{C}} =$$

$$= P(\psi_p'')\, \pi_{\psi p, \psi' p}\, g_{\mathcal{C}}\, u_{\mathcal{C}}^* = \dots = \pi_p\, P'(\psi')\, u_{\mathcal{C}}^* .$$

It follows that $P'(\psi)\, u_{\mathcal{C}}^* = P'(\psi')\, u_{\mathcal{C}}^*$, which completes the proof.

It follows from the lemma that $f_{\mathcal{D}}\, P'(\psi)\, u_{\mathcal{C}}^* = P''(\psi)\, f_{\mathcal{C}}\, u_{\mathcal{C}}^* = P''(\psi)\, g_{\mathcal{C}}\, u_{\mathcal{C}}^* = g_{\mathcal{D}}\, P'(\psi)\, u_{\mathcal{C}}^*$; therefore there exists a morphism $E_{\mathcal{C}\mathcal{D}} : E_{\mathcal{C}}(U) \longrightarrow E_{\mathcal{D}}(U)$ unique such that $P'(\psi)\, u_{\mathcal{C}}^* = u_{\mathcal{D}}^*\, E_{\mathcal{C}\mathcal{D}}(U)$, i.e. the following diagram commutes:

$$(1) \qquad \begin{array}{ccccc} E_{\mathcal{C}}(U) & \rightarrowtail & P(\mathcal{C}) & \rightrightarrows & P(\mathcal{C} * \mathcal{C}) \\ {\scriptstyle E_{\mathcal{C}\mathcal{D}}(U)} \downarrow & & \downarrow {\scriptstyle P'(\psi)} & & \downarrow {\scriptstyle P''(\psi)} \\ E_{\mathcal{D}}(U) & \rightarrowtail & P(\mathcal{D}) & \rightrightarrows & P(\mathcal{D} * \mathcal{D}) \end{array}$$

Furthermore the last part of the lemma shows that $E_{\mathcal{C}\mathcal{D}}(U)$ depends only on $\mathcal{C}$ and $\mathcal{D}$ and not on the choice of ψ . □

In case $\mathcal{C} = \mathcal{D}$ we may choose for ψ the identity on I and for ψ_i the identity morphisms and then it is clear that $P'(\psi)$ is the identity and so is $E_{\mathcal{C}\mathcal{C}}(U)$. If also $\mathcal{C} \leq \mathcal{D} \leq \mathcal{E}$ in $\mathfrak{C}(U)$, with $\mathcal{E} = (\gamma_z)_{z \in K}$, and $\chi : K \longrightarrow J$, χ_z $(z \in K)$ are such that $\gamma_z = \beta_{\chi z} \chi_z$ for all z , then, with ψ as above, we can define $\omega = \psi\chi : K \longrightarrow I$ and $\omega_z = \psi_{\chi z} \chi_z$, and see that $\gamma_z = \alpha_{\omega z}\, \omega_z$ for all $z \in K$ (this is how we showed the transitivity of $\leq$); furthermore,

$$\pi_z\, P'(\omega) = P(\omega_z)\, \pi_{\omega z} = P(\chi_z) P(\psi_{\chi z})\, \pi_{\psi\chi z} =$$
$$= P(\chi_z)\, \pi_{\chi z}\, P'(\psi) = \pi_z\, P'(\chi)\, P'(\psi)$$

for all z , so that $P'(\omega) = P'(\chi) P'(\psi)$; it follows that $E_{\mathcal{C}\mathcal{E}}(U) = E_{\mathcal{D}\mathcal{E}}(U)\, E_{\mathcal{C}\mathcal{D}}(U)$. Hence we now have a direct system over $\mathfrak{C}(U)$.

(3) We now let $E(U) = \varinjlim E_{\mathcal{C}}(U)$; it comes with maps $p_{\mathcal{C}}(U) : E_{\mathcal{C}}(U) \longrightarrow E(U)$. We also have a morphism $P(U) \longrightarrow E(U)$; indeed, $\mathcal{C} \leq \mathcal{D}$ in $\mathfrak{C}(U)$ implies (keeping the same notation as before)

$\pi_p\, P'(\psi)\, u_{\mathfrak{C}} = P(\psi_p)P(\alpha_{\psi p}) = P(\beta_p) = \pi_p\, u_{\mathfrak{D}}$, whence $P'(\psi)\, u_{\mathfrak{C}} = u_{\mathfrak{D}}$ and $E_{\mathfrak{C}\mathfrak{D}}(U) c_{\mathfrak{C}}(U) = c_{\mathfrak{D}}(U)$; since $\mathfrak{S}(U)$ is directed, we conclude that $c(U) = p_{\mathfrak{C}}(U)\, c_{\mathfrak{C}}(U)$ does not depend on $\mathfrak{C}$.

Finally, E is made into a presheaf as follows. Let $\gamma : W \longrightarrow U \in \mathfrak{U}(X)$. Then $E(\gamma) : E(W) \longrightarrow E(U)$ is induced by the restriction maps of P in the following manner. For each $C = (\alpha_i)_{i\in I} \in \mathfrak{S}(U)$, let $\alpha_i \gamma_i' = \gamma \alpha_i'$ be pullbacks, so that $C * W = (\alpha_i')_{i\in I} \in \mathfrak{S}(W)$. We then have a morphism

$$h' = h_C'(\gamma) = \prod_{i\in I} P(\gamma_i') : P(C) \longrightarrow P(C * W) .$$

Also, let $\alpha_j \xi_{jk} = \alpha_k \eta_{jk}$, $\alpha_j' \xi_{jk}' = \alpha_k' \eta_{jk}'$ be pullbacks. We then have, for each $j,k \in I$, a morphism γ_{jk}'' induced by the γ's , such that $\gamma_j' \xi_{jk}' = \xi_{jk} \gamma_{jk}''$, $\gamma'_k\, \eta_{jk}' = \eta_{jk} \gamma_{jk}''$. This yields a morphism

$$h'' = h_C''(\gamma) = \prod_{j,k\in I} P(\gamma_{jk}'') : P(C * C) \longrightarrow P((C * W)*(C * W)) .$$

(By definition, $\pi_{jk} h'' = P(\gamma_{jk}'')\pi_{jk}$, $\pi_i h' = P(\gamma_i')\pi_i$.)

For each j,k,

$\pi_{jk}\, h''\, f_C = P(\gamma_{jk}'')P(\xi_{jk})\, \pi_j = P(\xi_{jk}')P(\gamma_j')\, \pi_j = \pi_{jk}\, f_{C*W}\, h'$, so that $h''\, f_C = f_{C*W}\, h'$. Similarly, $h''\, g_C = g_{C*W}\, h'$. Therefore $f_{C*W}\, h'\, u_C^* = h''\, f_C\, u_C^* = h''\, g_C\, u_C^* = g_{C*W}\, h'\, u_C^*$ and there exists a morphism $E_C(\gamma) : E_C(U) \longrightarrow E_{C*W}(W)$ induced on equalizers, such that the following diagram commutes:

$$\begin{array}{ccccc} E_C(U) & \rightarrowtail & P(C) & \rightrightarrows & P(C * C) \\ \downarrow{\scriptstyle E_C(\gamma)} & & \downarrow{\scriptstyle h'} & & \downarrow{\scriptstyle h''} \\ E_{C*W}(W) & \rightarrowtail & P(C*W) & \rightrightarrows & P((C*W)*(C*W)) \end{array} \qquad (2)$$

Now assume that $C \leq \mathfrak{D}$ in $\mathfrak{S}(U)$. With the notation as before, we have a mapping $\psi : J \longrightarrow I$ and morphisms ψ_p such that $\beta_p = \alpha_{\psi p}\psi_p$

for all p. We also have pullbacks $\beta_p \gamma'_p = \gamma \beta'_p$ yielding a covering $\mathcal{D} * W = (\beta'_p)_{p \in J} \in \mathfrak{C}(W)$, and maps $h'_{\mathcal{D}}(\gamma) : P(\mathcal{D}) \longrightarrow P(\mathcal{D}*W)$, $E_{\mathcal{D}}(\gamma) : E_{\mathcal{D}}(U) \longrightarrow E_{\mathcal{D}*W}(W)$. From $\alpha_{\psi p} \psi_p \gamma'_p = \gamma \beta'_p$ we also obtain for each morphism p a morphism ψ'_p such that $\psi_p \gamma'_p = \gamma'_{\psi p} \psi'_p$ and $\beta'_p = \alpha'_{\psi p} \psi'_p$; in particular this shows that $C*W \leq \mathcal{D}*W$, and yields a map $P'(\psi') : P(C*W) \longrightarrow P(\mathcal{D}*W)$. Now for each p,

$$\pi_p \, P'(\psi') \, h'_C = P(\psi')P(\gamma'_{\psi p}) \, \pi_{\psi p} = P(\gamma'_p)P(\psi_p) \, \pi_{\psi p} = \pi_p \, h'_{\mathcal{D}} \, P'(\psi) \quad ,$$

which shows that $P'(\psi') \, h'_C = h'_{\mathcal{D}} \, P'(\psi)$. Thus every face of the following diagram

(3)

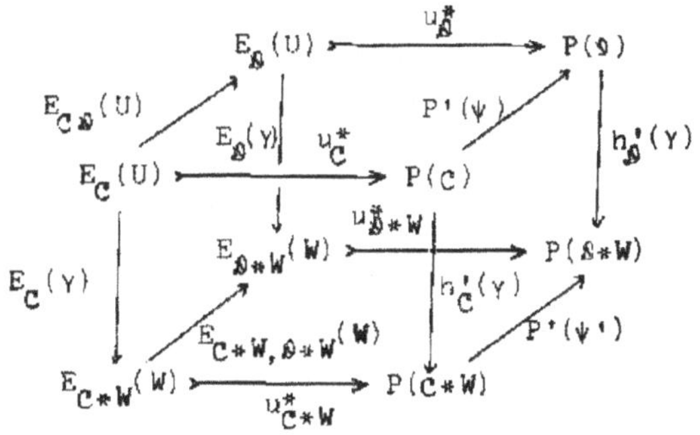

commutes except perhaps for the left face. But then this face commutes too, since $u^*_{\mathcal{D}*W}$ is a monomorphism; i.e. $E_{\mathcal{D}}(\gamma)E_{C\mathcal{D}}(U) = E_{C*W,\mathcal{D}*W}(W)E_C(\gamma)$.

It follows that $p_{C*W}(W) \, E_C(\gamma) = p_{\mathcal{D}*W}(W) \, E_{\mathcal{D}}(\gamma) \, E_{C\mathcal{D}}(U)$, i.e. we have shown that $(p_{C*W}(W) \, E_C(\gamma))_{C \in \mathfrak{C}(U)}$ is a cocompatible family; hence it induces a morphism $E(\gamma) : E(U) \longrightarrow E(W)$, unique such that the following diagram commutes:

(4)

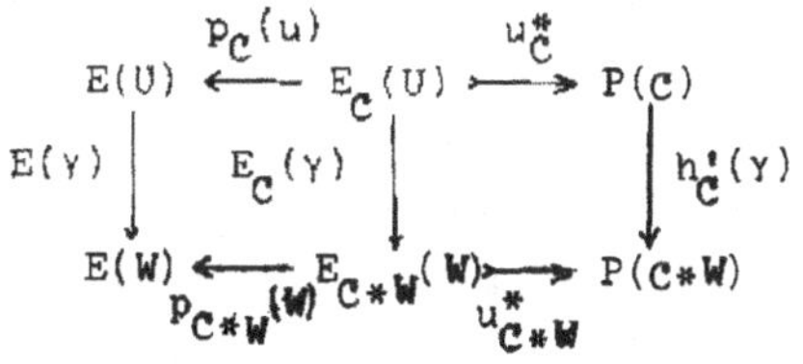

$\gamma = 1_U$ then $\alpha_i 1 = \gamma\alpha_i$ is a pullback for every i, in other words $C * W = C$ and $h' = 1$, for every $C \in \mathfrak{C}(U)$; hence $E_C(1_U) = 1$ and (going to the colimit) $E(1_U) = 1_{E(U)}$. If γ is arbitrary and $\delta : Z \longrightarrow W \in \mathfrak{U}(X)$, then to construct $E(\gamma)$, $E(\delta)$, we take pullbacks $\gamma\alpha'_i = \alpha_i\gamma'_i$, $\delta\alpha''_i = \alpha'_i\delta'_i$ for each $C = (\alpha_i)_{i\in I} \in \mathfrak{C}(U)$; juxtaposition yields pullbacks $(\gamma\delta)\alpha''_i = \alpha_i(\gamma'_i\delta'_i)$, which means that $\gamma'_i\delta'_i = (\gamma\delta)'_i$ and $(C * W) * Z = C * Z$. Hence all faces of the following diagram commute except perhaps the bottom face:

(5)

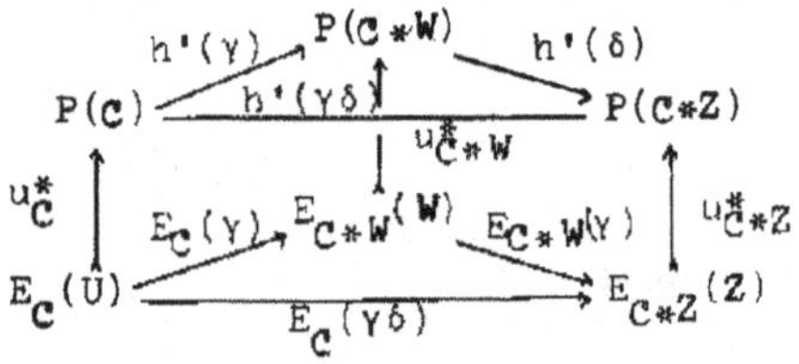

then the bottom face commute anyway, since u^*_{C*Z} is a monomorphism. This shows that $E_C(\gamma\delta) = E_{C*W}(\delta)\, E_C(\gamma)$. Hence

$$E_C(\gamma\delta)\; p_{C(U)} = p_{C*Z}\; E_C(\gamma\delta) = p_{C*Z}(U)\; E_{C*W}(\delta)\; E_C(\gamma) =$$
$$= E(\delta)\; p_{C*W}(U)\; E_C(\gamma) = E(\delta)\; E(\gamma)\; p_{C(U)}$$

for all C, and $E(\gamma\delta) = E(\delta)E(\gamma)$. Therefore E is indeed a presheaf. We state this with two other properties of E:

4. Proposition 2.2. Let $\mathfrak{a}$ be a complete category having directed colimits, and X be any Grothendieck topology. Then for each $P \in \mathcal{P}(X,\mathfrak{a})$ the above construction yields a presheaf E and a morphism $c : P \longrightarrow E$ such that every morphism a of P to a sheaf factors uniquely through c ($a = tc$ for some t).

Proof. We already know that E is a presheaf, and have morphisms $c(U) : P(U) \longrightarrow E(U)$ for each $U \in \mathfrak{U}(X)$. Let $\gamma : W \longrightarrow U \in \mathfrak{U}(X)$. Keeping the notation as before, we have for each $C \in \mathfrak{C}(U)$ the following diagram, in which, by definition of the various maps under consideration, every triangle and square commutes except possibly for the three squares fanning out of $P(\gamma)$:

(6)

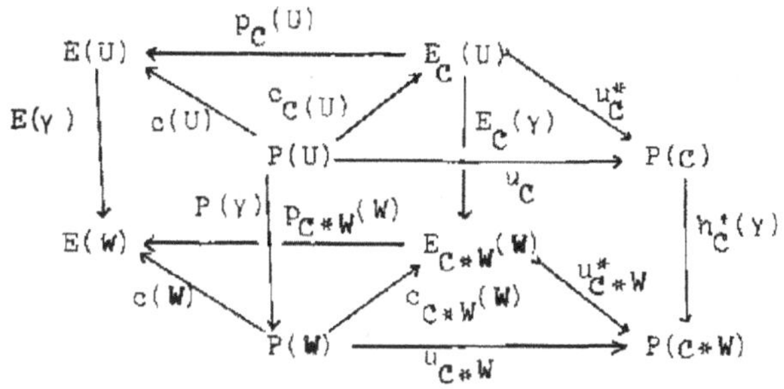

Now, for each $i \in I$,

$$\pi_i h' u_C = P(\gamma'_i)P(\alpha_i) = P(\alpha'_i)P(\gamma) = \pi_i u_{C*W} P(\gamma) \quad ,$$

so that the front square commutes. Since u^*_{C*W} is a monomorphism, it follows that the diagonal square also commutes; and then the left square commutes. This shows that $c = (c(U))_{U \in \mathfrak{U}(X)}$ is a morphism of presheaves. □

We now let $a : P \longrightarrow F$ be a morphism from P to a sheaf F. Take $U \in \mathfrak{U}(X)$, $C \in \mathfrak{C}(U)$. From $C = (\alpha_i)_{i \in I}$, $\alpha_i : U_i \longrightarrow U$ we obtain a diagram:

(7)

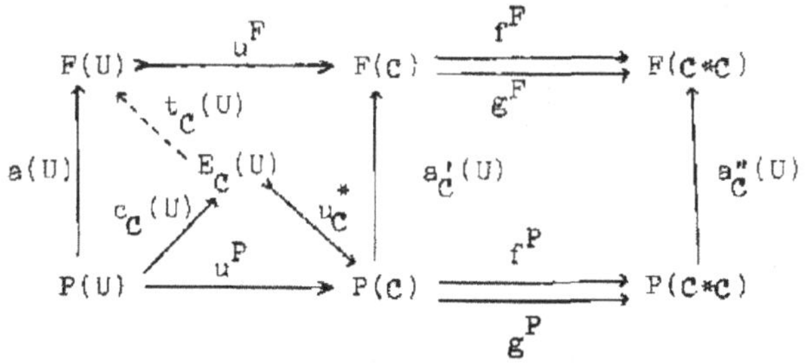

where $a'_C(U) = a' = \prod_{i \in I} a(U_i)$, $a''_C(U) = a'' = \prod_{j,k \in I} a(U_j * U_k)$ and $t_C(U)$ will be presently constructed. For each j,k,

$$\pi_{jk} f^F a' = F(\xi_{jk}) a(U_j) \pi_j = a(U_j * U_k) P(\xi_{jk}) \pi_j = \pi_{jk} a'' f^P \quad ,$$

so that $f^F a' = a'' f^P$. Similarly, $g^F a' = a'' g^P$. Since F is a sheaf, there is a morphism $t_C(U) : E_C(U) \longrightarrow F(U)$ (induced on equalizers) such that $a' u^*_C = u^F_C \, t_C(U)$; since u^F is a monomorphism, we also have $t_C(U) c_C(U) = a(U)$.

If $C \leq \mathfrak{D}$ in $\mathfrak{C}(U)$, then we obtain a diagram

(8)

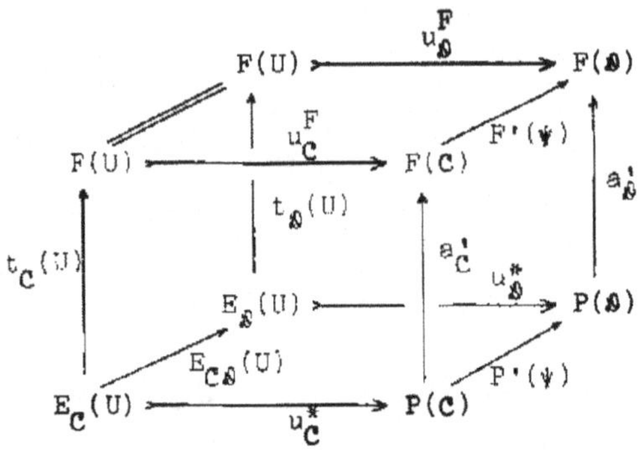

in which all squares commute except perhaps for the left and right squares. With the notation as above,

$$\pi_p \, a'_{\mathfrak{D}} \, P'(\psi) = a(V_p)P(\psi_p)\pi_{\psi p} = F(\psi_p)a(U_{\psi p})\pi_{\psi p} = \pi_p F'(\psi) a'_C$$

for all p, so that the square on the right commutes also. Since $u^F_{\mathfrak{D}}$ is a monomorphism, it follows that the square on the left commutes, i.e. $t_{\mathfrak{D}}(U) \, E_{C\mathfrak{D}}(U) = t_C(U)$.

In other words, $(t_C(U))_{C \in \mathfrak{C}(U)}$ is a cocompatible family. Therefore it induces a morphism $t(U) : E(U) \longrightarrow F(U)$, such that $t_C(U) = t(U) \; p_C(U)$ for all C. Note that

$$t(U)c(U) = t(U) \; p_C(U) \; c_C(U) = t_C(U) \; c_C(U) = a(U) .$$

To prove the existence of the factorization, we only have to show that <u>t is a morphism of presheaves</u>.

Let $\gamma : W \longrightarrow U \in \mathfrak{U}(X)$. We have induced morphisms $h'^P(\gamma) : P(C) \longrightarrow P(C * W)$, $h'^F(\gamma) : F(C) \longrightarrow F(C * W)$. For each i ,

$$\pi_i h'^F a'_C = F(\gamma'_i)a(U_i)\pi_i = a(U_i * W)P(\gamma'_i)\pi_i = \pi_i a'_{C*W}\, h'^P$$

so that $h'^F a'_C = a'_{C*W} h'^P$. Similarly, $u^F_{C*W}\, F(\gamma) = h'^F\, u^F_C$. Since diagrams (4) and (7) commute, this yields a diagram

(9)

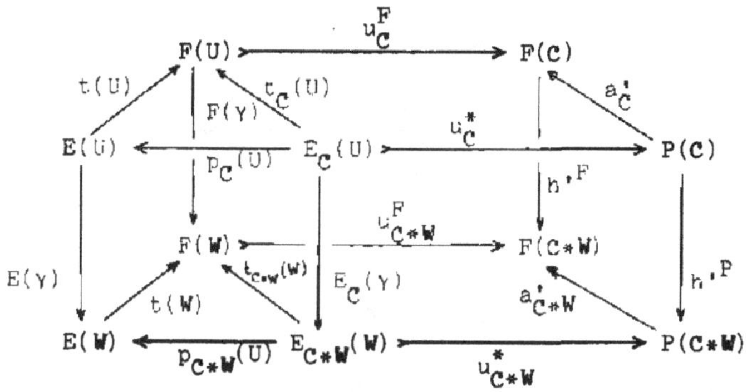

which commutes except perhaps for the two squares that connect $F(\gamma)$ to $E(\gamma)$ and $E_C(\gamma)$. First we see that the latter does commute, since u^F_{C*W} is a monomorphism. This yields $t(W)E(\gamma)p_C(U) = F(\gamma)t(U)p_C(U)$; since this is true for all C , it follows that $t(W)E(\gamma) = F(\gamma)t(U)$, i.e. the diagram commutes, and we have shown that t is a morphism of presheaves. ○

The uniqueness of the factorization will follow from a description of u^E_C , namely $u^E_C\, p_C(U) = c'_C(U)\, u^*_C$ for all $C \in \mathfrak{C}(U)$. (4)

To see that, set, as usual, $C = (\alpha_i)_{i\in I}$, $\alpha_i : U_i \longrightarrow U$; we remember that $E(\alpha_i)$ is induced by $E_C(\alpha_i) : E_C(U) \longrightarrow E_{C*U_i}(U_i)$, which is in turn induced by $h' : P(C) \longrightarrow P(C*U_i)$; the pullbacks we need to describe h' are precisely $\alpha_i \xi_{ij} = \alpha_j \eta_{ij}$, so that $C * U_i = (\xi_{ij})_{j\in I}$ and $h' = \prod_{j\in I} P(\eta_{ij})$. We now consider the following diagram, in which the outer square and bottom triangle are already known to commute:

(10)

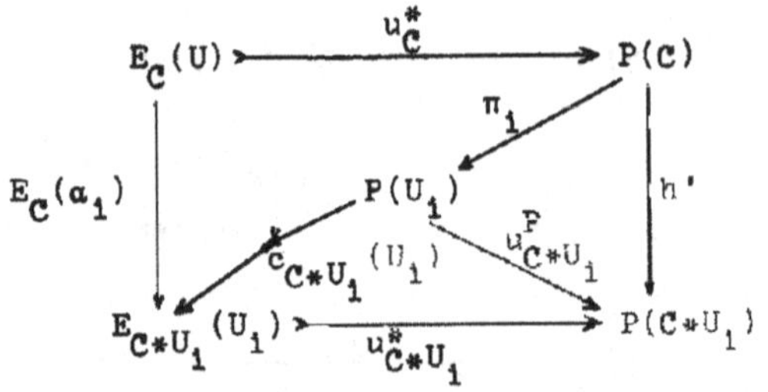

Note that the triangle on the right need not commute; however, for each $j \in I$:

$$\pi_j u^P_{C*U_i} \pi_i u^*_C = P(\xi_{ij}) \pi_i u^*_C = \pi_{ij} f^P_C u^*_C ,$$

$$\pi_j h' u^*_C = P(\eta_{ij}) \pi_j u^*_C = \pi_{ij} g^P_C u^*_C ,$$

and it follows that $u^P_{C*U_i} \pi_i u^*_C = h' u^*_C$. Since $u^*_{C*U_i}$ is a monomorphism, it follows that the upper left triangle commutes, i.e. $E_C(\alpha_i) = c_{C*U_i} \pi_i u^*_C$. Hence

$$c(U_i) \pi_i u^*_C = p_{C*U_i} c_{C*U_i} \pi_i u^*_C = p_{C*U_i} E_C(\alpha_i) = E(\alpha_i) p_C .$$

Applying $\underset{i \in I}{X}$ to both sides yields the desired formula $c'_C(U) u^*_C = u^E_C p_C(U)$, i.e. the following diagram commutes:

(11)

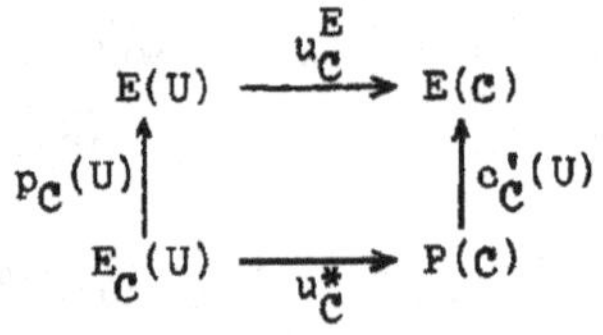

Let now $t_1, t_2 : E \longrightarrow F$ be morphisms such that $t_1 c = t_2 c$, where F is a sheaf (in fact, the uniqueness still holds if F is only a monopresheaf). For each $C \in \mathfrak{C}(U)$, we have $t_1' c' = t_2' c'$ and hence

$$u^F_C t_1(U) p_C(U) = t_1' u^E_C p_C(U) = t_1' c' u^*_C =$$

$$= t_2' c' u^*_C = \dots = u^F_C t_2(U) p_C(U) ;$$

since this holds for all C and u^F is a monomorphism, it follows that $t_1(U) = t_2(U)$, whence $t_1 = t_2$; this completes the proof. □

Corollary 2.3. Further assume that directed colimits in $\mathcal{G}$ preserve (pointwise) monomorphisms. If in 2.2 a is a pointwise monomorphism, then so is t.

Proof. Then $a'_{\mathbf{C}}(U)$ is a monomorphism for all $\mathbf{C} \in \mathfrak{C}(U)$; hence it is clear on diagram (7) that $t_{\mathbf{C}}(U)$ is also a monomorphism. By the hypothesis, so is $t(U)$.

3. THE CASE OF A C_4 REGULAR CATEGORY.

1. We now assume that $\mathcal{G}$ is a C_4 regular category.

Lemma 3.1. For any presheaf P, E is a monopresheaf.

Proof. Take $\mathbf{C} \in \mathfrak{C}(U)$ $[\mathbf{C} = (\alpha_i)_{i \in I}, \alpha_i : U_i \longrightarrow U]$; we want to show that $u^E_{\mathbf{C}}$ is a monomorphism. Now $E(\mathbf{C}) = \prod_{i \in I} \varinjlim E_{\mathbf{C}_i}(U_i)$ and we can apply theorem II.4.4, which in this situation says that the morphisms $p'_\tau = \prod_{i \in I} p_{\tau i}(U_i)$, $\tau \in T = \prod_{i \in I} \mathfrak{C}(U_i)$, induce a monomorphism

$$t : \varinjlim_{\tau \in T} \prod_{i \in I} E_{\tau i}(U_i) \longrightarrow E(\mathbf{C}) .$$

For each $\tau \in T$, put $\tau i = (\beta_p)_{p \in J_i}$, where we assume that the sets J_i are pairwise disjoint and disjoint from I ; then $\mathbf{C}_\tau = (\alpha_i \beta_p)_{p \in J_i, i \in I}$ is in $\mathfrak{C}(U)$. We also put $\bigcup_{i \in I} J_i = J$ and $\alpha_i \beta_p = \gamma_p : V_p \longrightarrow U$. We now interrupt the proof to observe:

Lemma 3.2. For every $\mathbf{C} \in \mathfrak{C}(U)$, $\{\mathbf{C}_\tau ; \tau \in T\}$ is a cofinal subset of $\mathfrak{C}(U)$.

Proof of 3.2. Take $\mathbf{D} \in \mathfrak{C}(U)$. Define $\tau \in T$ by: $\tau i = \mathbf{D} * U_i \in \mathfrak{C}(U_i)$.

Then $C_\tau = \mathcal{D} * C$, so that $\mathcal{D} \le C_\tau$.

We now resume the proof of 3.1. Let π_o be the projection

$$\pi_o : P(C_\tau * C_\tau) = \prod_{p,q\in J} P(V_p * V_q) \longrightarrow \prod_{i\in I} \prod_{p,q\in J_i} P(V_p * V_q) = \prod_{i\in I} P(\tau i * \tau i)$$

(note that $\bigcup_{i\in I} J_i \pi J_i \subseteq J \pi J$). The pullbacks $\gamma_p \xi_{pq} = \gamma_q \eta_{pq}$ used in evaluating $\tau i * \tau i$ also serve for $C_\tau * C_\tau$; hence $\pi_{pq} \pi_o f^P_{C_\tau} = P(\xi_{pq}) = \pi_{pq} f^P_{\tau i}$ for all $p,q \in J_i$, whence $\pi_o f^P_{C_\tau} = \prod_{i\in I} f^P_{\tau i}$. Thus we have a commutative diagram (where u_τ will presently be constructed):

$$\begin{array}{ccccc}
\prod_{i\in I} E_{\tau i}(U_i) & \xrightarrow{\prod_{i\in I} u^*_{\tau i}} & \prod_{i\in I} P(\tau i) & \xrightarrow{\prod_{i\in I} f^P_{\tau i}} & \prod_{i\in I} P(\tau i * \tau i) \\
\uparrow u_\tau & & \| & & \uparrow \pi_o \\
E_{C_\tau}(U_i) & \xrightarrow[u^*_{C_\tau}]{} & P(C_\tau) & \xrightarrow[f^P_{C_\tau}]{} & P(C_\tau * C_\tau)
\end{array} \tag{12}$$

There is a similar commutative diagram with g's instead of f's . Now, products preserve equalizers, and hence there is a morphism u_τ induced on equalizers, such that $u^*_{C_\tau} = (\prod_{i\in I} u^*_{\tau i})\, u_\tau$. Note that u_τ is a monomorphism.

We now prove that $p'_\tau u_\tau = u^E_C\, p_{C_\tau}$. First, note that $C_\tau * U_i \le \tau i$; more precisely, take pullbacks $\gamma_p \alpha'_{pi} = \alpha_i \gamma'_{ip}$ $(p \in J)$, so that $C_\tau * U_i = (\gamma'_{ip})_{p\in J}$; let $\psi : J_i \longrightarrow J$ be the inclusion ; if $p \in J_i$ (so that $\psi p = p$), then $\gamma_p 1 = \alpha_i \beta_p$ and the pullback yield a morphism ψ_p such that $\alpha'_{pi} \psi_p = 1$ and $\beta_p = \gamma'_{ip} \psi_p$. We now have a three-dimensional diagram (next page) in which, of the six areas indicated, area ⑥ commutes trivially, and areas ②, ⑤ and ③ commute by definition of the various E maps therein (see diagrams (4) and (1)). For each $p \in J_i$,

$$\pi_p\, P'(\psi)\, h' = P(\psi_p) P(\alpha'_{pi})\, \pi_p = \pi_p \quad ,$$

and it follows that area ① also commutes:

(13)

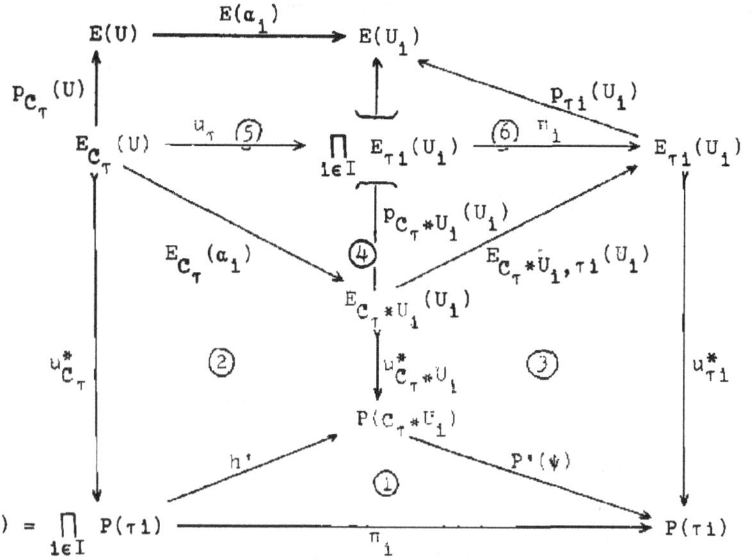

Finally, $u^*_{\tau i} \pi_i u_\tau = \pi_i u^*_{C_\tau}$ by definition of u_τ; since $u^*_{\tau i}$ is a monomorphism, it follows that area ④ commutes. Hence the whole diagram is commutative, in particular

$$p_{\tau i}(U_1) \pi_i u_\tau = E(\alpha_1) p_{C_\tau}(U_1) .$$

Applying $\underset{i \in I}{\times}$ to both sides yields the desired formula $p'_\tau u_\tau = u^E_C p_{C_\tau}$.

We now take directed colimits (over T). In view of 3.2, this sends the commutative square below left to the commutative square below right:

(14)

$$
\begin{array}{ccc}
E(U) & \xrightarrow{u^E_C} & E(C) \\
{\scriptstyle p_{C_\tau}}\big\uparrow & & \big\uparrow{\scriptstyle p'_\tau} \\
E_{C_\tau}(U) & \xrightarrow{u_\tau} & \underset{i \in I}{\prod} E_{\tau i}(U_1)
\end{array}
\qquad
\begin{array}{ccc}
E(U) & \xrightarrow{u^E_C} & E(C) \\
\big\| & & \big\uparrow{\scriptstyle t} \\
E(U) & \xrightarrow{u} & \underset{\tau \in T}{\varinjlim} \underset{i \in I}{\prod} E_{\tau i}(U_1)
\end{array}
$$

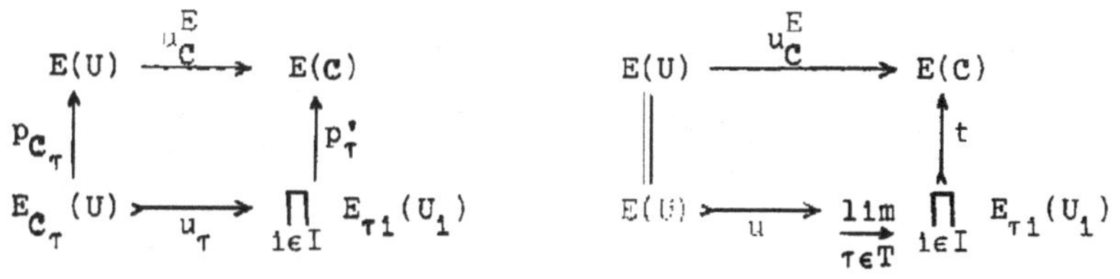

where t is a monomorphism and so is $u = \varinjlim u_\tau$. Then u^E_C is a monomorphism, q.e.d.

Lemma 3.3. If P is a monopresheaf, then E is a sheaf.

Proof. When P is a monopresheaf, $u_C^P = u_C^* \, c_C(U)$ shows that $c_C(U)$ is a monomorphism; hence $c(U) = \varinjlim c_C(U)$ is also a monomorphism. In other words, $c : P \longrightarrow E$ is a monomorphism. Then looking at the commutative diagram (11), where $c_C'(U)$ and u_C^* are monomorphisms, shows that $p_C(U)$ is a monomorphism. This shows that every direct system $E\_(U) : \mathfrak{C}(U) \longrightarrow \mathfrak{a}$ is monic.

We now start the proof as for lemma 3.1; this time, by the above, theorem II.4.4 tells us that t is an isomorphism.

Given $j, k \in I$, certain relations exist between the coverings of U_j, U_k we already have and coverings of $U_j * U_k$ that arise from these. For each $p \in J_j$, $q \in J_k$, consider the diagram in $\mathfrak{U}(X)$:

(15)

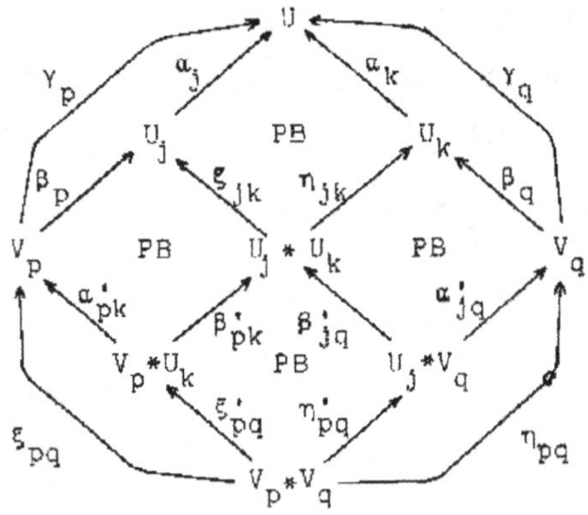

in which each square is a pullback, arranged so that juxtaposition yields the previous pullbacks $\gamma_p \alpha'_{pk} = \alpha_k \gamma'_{pk}$, $\alpha_j \gamma'_{jq} = \gamma_q \alpha'_{jq}$, $\gamma_p \xi_{pq} = \gamma_q \eta_{pq}$. We remember that $\tau j = (\beta_p)_{p \in J_j}$, $\tau k = (\beta_q)_{q \in J_k}$; the diagram yields a covering $\tau j * (U_j * U_k) = (\beta'_{pk})_{p \in J_j}$, for which we shall abuse the notation by calling it $\tau j * U_k$. Similarly we obtain

a covering $\tau k * U_j\,[= \tau k * (U_j * U_k)] = (\beta'_{jq})_{q\in J_k}$ also in $\mathfrak{C}(U_j * U_k)$. The diagram finally yields a covering $\tau j * \tau k\ [= (\tau j * (U_j * U_k)) * (\tau k * (U_j * U_k))]$ which refines both $\tau j * U_k$ and $\tau k * U_j$; it is given by:
$\tau j * \tau k = (\gamma_{pq})_{p\in J_j,\, q\in J_k}$, where $\gamma_{pq} = \beta'_{pk}\xi'_{pq} = \beta'_{jq}\eta'_{pq}$; it is this finer covering we need for the proof. That $\tau j * \tau k \geq \tau j * U_k$ is seen more precisely by considering the projection $\chi : J_j \sqcap J_k \longrightarrow J_j$ and morphisms $\chi_{pq} = \xi'_{pq}$.

Then we have a commutative diagram:

(16)

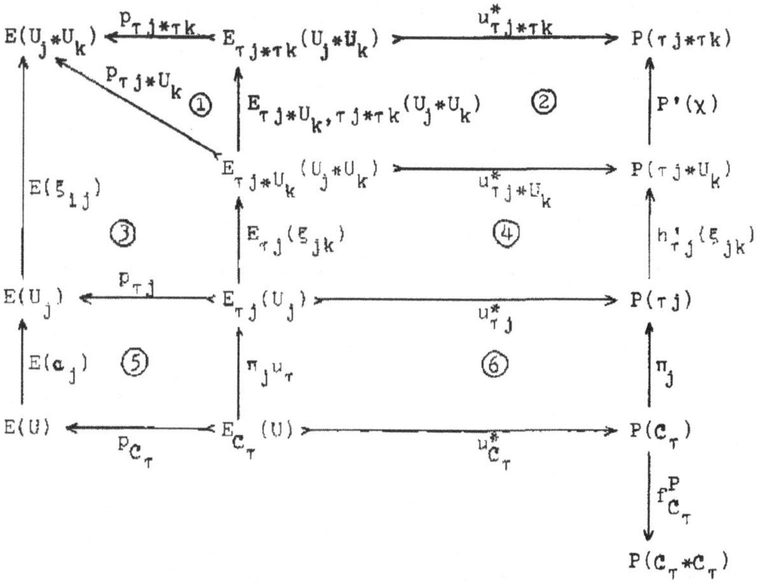

where area ① commutes trivially, areas ②, ③, ④ commute by definition of the E maps therein (see diagrams (1) and (4), and areas ⑤ and ⑥ commute because diagram (13) commutes.

We now merge areas ① - ③, areas ② - ④ , and take products (over j,k for the top row, over j for the middle row). This yields the next commutative diagram, in which $\bar{u} = \underset{i\in I}{\times} u^*_{\tau i}$, $p'_\tau = \prod_{i\in I} p_{\tau i}$ (as defined before), $p''_\tau = \prod_{j,k\in I} p_{\tau j * \tau k}$ and $\bar{\bar{u}} = \prod_{j,k\in I} u^*_{\tau j * \tau k}$.

All four are monomorphisms (p' and p'' because in each monic direct system $E\_(\mathbf{W}) : \mathfrak{C}(\mathbf{W}) \longrightarrow \mathfrak{a}$ the morphisms p are monomorphisms, as observed at the beginning of the proof). In addition, using the definition of $P'(\chi)$ and $h'_{\tau j}(\xi_{jk})$, we see on diagram (15) that

$$\pi_{pq}\ P'(\chi)h'_{\tau j}(\xi_{jk}) = P(\xi'_{pq})P(\alpha'_{pk})\pi_p = P(\xi_{pq})\pi_p = \pi_{pq}\ f^P_{\mathbf{C}_\tau}$$

for all $p \in J_j$, $q \in J_k$; therefore the vertical map (top right) on the new diagram (17) below is $f^P_{\mathbf{C}_\tau}$. The diagram:

$$(17)\qquad \begin{array}{ccccc}
E(\mathbf{C}*\mathbf{C}) & \xleftarrow{\;p''_\tau\;} & \prod_{j,k\in I} E_{\tau j*\tau k}(U_j*U_k) & \xrightarrow{\;\bar{\bar{u}}\;} & \prod_{j,k\in I} P(\tau j*\tau k) \\
\uparrow{\scriptstyle f^E_{\mathbf{C}}} & & \uparrow & & \uparrow{\scriptstyle f^P_{\mathbf{C}_\tau}} \\
E(\mathbf{C}) & \xleftarrow{\;p'_\tau\;} & \prod_{i\in I} E_{\tau i}(U_i) & \xrightarrow{\;\bar{u}\;} & P(\mathbf{C}_\tau) \\
\uparrow{\scriptstyle u^E_{\mathbf{C}}} & & \uparrow{\scriptstyle u_\tau} & & \| \\
E(U) & \xleftarrow{\;p_{\mathbf{C}_\tau}\;} & E_{\mathbf{C}_\tau}(U) & \xrightarrow{\;u^*_{\mathbf{C}_\tau}\;} & P(\mathbf{C}_\tau) \\
 & & & & \downarrow{\scriptstyle f^P_{\mathbf{C}_\tau}} \\
 & & & & \prod_{j,k\in I} P(\tau j*\tau k) = P(\mathbf{C}_\tau*\mathbf{C}_\tau)
\end{array}$$

Since we went up in the coverings to $\tau j * \tau k$, which is "symmetric in j and k", we obtain, by working on the other side of diagram (15), a commutative diagram which is the same as (17) but with g's instead of f's (and a different unnamed morphism in the middle column); all other morphisms will remain unchanged. Now p''_τ and $\bar{\bar{u}}$ are monomorphisms and when we consider both diagrams it is evident that $\mathrm{Equ}(f^E_{\mathbf{C}}p'_\tau, g^E_{\mathbf{C}}p'_\tau) = \mathrm{Equ}(f^P_{\mathbf{C}_\tau}\bar{u}, g^P_{\mathbf{C}_\tau}\bar{u})$.

This in turn implies that $\mathrm{Equ}(f^E_{\mathbf{C}}p'_\tau, g^E_{\mathbf{C}}p'_\tau) = \mathrm{Im}\ u_\tau$. To see this, we first note that $u^E_{\mathbf{C}}$ is a monomorphism, by 3.1, and the diagram then shows that u_τ is a monomorphism. Next,

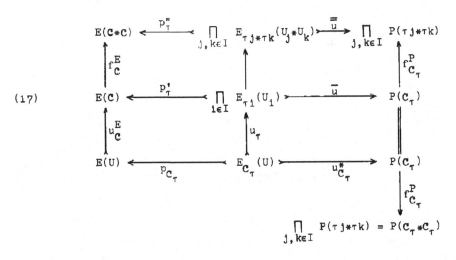

$$f^P_{\mathbf{C}_\tau} \bar{u}\, u_\tau = f^P_{\mathbf{C}_\tau} u^*_{\mathbf{C}_\tau} = g^P_{\mathbf{C}_\tau} u^*_{\mathbf{C}_\tau} = g^P_{\mathbf{C}_\tau} \bar{u}\, u_\tau \quad .$$

Finally, $f^P_{\mathbf{C}_\tau} \bar{u}\, a = g^P_{\mathbf{C}_\tau} \bar{u}\, a$ implies $\bar{u}\, a = u^*_{\mathbf{C}_\tau} x = \bar{u}\, u_\tau\, x$ for some x, whence $a = u_\tau x$ for some x. Hence $u_\tau \in \mathrm{Equ}(f^P_{\mathbf{C}_\tau} \bar{u}, g^P_{\mathbf{C}_\tau} \bar{u}) = \mathrm{Equ}(f^E_{\mathbf{C}} p'_\tau, g^E_{\mathbf{C}} p'_\tau)$

If now we go to the directed colimit over T, then we saw at the beginning of the proof that the morphisms p'_τ induce an isomorphism; so do the $p_{\mathbf{C}_\tau}$, by 3.2; hence it follows from II.4.3 that

$$\begin{aligned} \mathrm{Equ}(f^E_{\mathbf{C}}, g^E_{\mathbf{C}}) &= \bigvee_{\tau\in T} (p'_\tau)_s\, \mathrm{Equ}\,(\, f^E_{\mathbf{C}} p'_\tau\,,\, g^E_{\mathbf{C}} p'_\tau\,) = \\ &= \bigvee_{\tau\in T} (p'_\tau)_s\, \mathrm{Im}\, u_\tau = \bigvee_{\tau\in T} \mathrm{Im}\, p'_\tau u_\tau = \bigvee_{\tau\in T} \mathrm{Im}\, u^E_{\mathbf{C}} p_{\mathbf{C}_\tau} = \\ &= (u^E_{\mathbf{C}})_s \bigvee_{\tau\in T} \mathrm{Im}\, p_{\mathbf{C}_\tau} = \mathrm{Im}\, u^E_{\mathbf{C}} \quad . \end{aligned}$$

Since $u^E_{\mathbf{C}}$ is a monomorphism, this proves that E is a sheaf.

Hence we have proved:

<u>Theorem</u> 3.4. If $\mathcal{G}$ is a C_4 regular category, then for any Grothendieck topology X, $\mathcal{F}(X,\mathcal{G})$ is coreflective in $\mathcal{P}(X,\mathcal{G})$, and Heller and Rowe's construction yields the coreflection in at most two steps.

2. We denote by $\hat{}\ : \mathcal{P}(X,\mathcal{G}) \longrightarrow \mathcal{F}(X,\mathcal{G})$ the coreflection. In order to obtain a well-defined functor, we take the functor obtained by applying twice the clearly functorial Heller and Rowe construction, and amend it (i.e. change it by a natural isomorphism) so that $\hat{P}$ is obtained in one step from P if P is a monopresheaf, and $\hat{P} = P$ if P is a sheaf [no such fuss is necessary with a stronger set theory].

Since $\mathcal{G}$ is C_4, 2.3 holds, so that $\hat{}$ preserves monomorphisms when their codomain is a sheaf. We use this to prove:

<u>Theorem</u> 3.5. If $\mathcal{G}$ is a C_4 regular category, then for any Grothendieck topology X, $\mathcal{F}(X,\mathcal{G})$ is a C_3 regular category.

<u>Proof</u>. First the functor category $\mathcal{P}(X,\mathcal{G})$ is regular by I.2.1,

with pointwise regular decompositions. It is also complete and cocomplete, like $\mathcal{G}$, and since everything in $\mathcal{P}(X,\mathcal{G})$ works pointwise, including subobjects and their intersections and unions, $\mathcal{P}(X,\mathcal{G})$ is in fact a C_4 regular category. The coreflective subcategory $\mathfrak{F}(X,\mathcal{G})$ inherits completeness and cocompleteness from $\mathcal{P}(X,\mathcal{G})$.

The existence of regular decompositions in $\mathfrak{F}(X,\mathcal{G})$ then follows from I.1.6. However, it is interesting (and necessary) to see what they look like. First monomorphisms in the finitely complete category $\mathfrak{F}(X,\mathcal{G})$ can be characterized by their kernel pairs, which are the same in $\mathfrak{F} = \mathfrak{F}(X,\mathcal{G})$ as in $\mathcal{P} = \mathcal{P}(X,\mathcal{G})$, and it follows that the monomorphisms of $\mathfrak{F}$ coincide with the pointwise monomorphisms of $\mathfrak{F}$. The regular epimorphisms are given by:

<u>Proposition</u> 3.6. Let $f \in \mathfrak{F}$ have the regular decomposition (m,p) in $\mathcal{P}$. Then f is a regular epimorphism (in $\mathfrak{F}$) if and only if $\hat{m}$ is an isomorphism.

<u>Proof</u> of 3.6. First, assume that $\hat{m}$ is an isomorphism. Then $a,b \in \mathfrak{F}$, $af = bf$ implies $am = bm$, $a\hat{m} = \hat{a}\hat{m} = \hat{b}\hat{m} = b\hat{m}$ and $a = b$, so that f is an epimorphism in $\mathfrak{F}$. Now let $fx = fy$ be a pullback and $g \in \mathfrak{F}$ be such that $gx = gy$. Since, in $\mathcal{P}$, $px = py$ is also a pullback (as m is a monomorphism) and p is a regular epimorphism, we have $g = tp$ for some t. Then also $g = \hat{t}\hat{p}$ and since $\hat{m}$ is an isomorphism g factors through f, uniquely since f is an epimorphism. This shows that $f \in \mathrm{Coequ}_{\mathfrak{F}}(x,y)$ and hence that f is a regular epimorphism.

Conversely assume that f is a regular epimorphism. Let M be the domain of m and $c : M \longrightarrow \hat{M}$ be the coreflection. Then $m = \hat{m}c$, $\hat{p} = cp$. Since $\hat{c} = 1$ is an isomorphism, and c is a monomorphism (since $m = \hat{m}c$, or for the more general reason that M is a monopresheaf), the first part of the proof shows that $\hat{p}$ is a regular epimorphism of $\mathfrak{F}$. On the other hand, $\hat{m}$ is a monomorphism, by 2.3. Hence

$(\hat{m},\hat{p})$ is a regular decomposition of f and since f is a regular epimorphism it follows that $\hat{m}$ is an isomorphism.

We now interrupt the proof of the theorem to show:

Proposition 3.7. Let $\mathcal{G}$ be a C_4 regular category, and X be any Grothendieck topology. Then the coreflection $\mathcal{P}(X,\mathcal{G}) \longrightarrow \mathcal{F}(X,\mathcal{G})$ is exact.

Proof. We already know that it preserves colimits. On the other hand, it is obtained by applying twice Heller and Rowe's construction which, being defined in terms of products, equalizers and directed colimits of $\mathcal{G}$, commutes with finite limits. It is therefore an exact functor. [This provides an alternate proof of 3.6 above.]

To prove that $\mathcal{F}(X,\mathcal{G})$ is regular, it now suffices to prove that it satisfies the pullback axiom. First we establish the following particular case:

Lemma 3.8. Let

$$\begin{array}{ccc} M & \xrightarrow{m} & G \\ {\scriptstyle r}\uparrow & & \uparrow{\scriptstyle q} \\ N & \xrightarrow[n]{} & K \end{array}$$

be a pullback in $\mathcal{P}(X,\mathcal{G})$, where G,K are sheaves and m, n monomorphisms. If $\hat{m}$ is an isomorphism, then so is $\hat{n}$.

Proof. Then M, N are monopresheaves and Heller and Rowe's construction gives their coreflections in one step; for M, it is given on the commutative diagram (18) below, obtained from (7), where the notation is as usual and we recall that $\hat{m}_C(U)$ is induced on equalizers (considering the similar diagram with g's instead of f's) and in turn induces $\hat{m}(U)$ when we take directed colimits over $\mathfrak{C}(U)$:

(18)

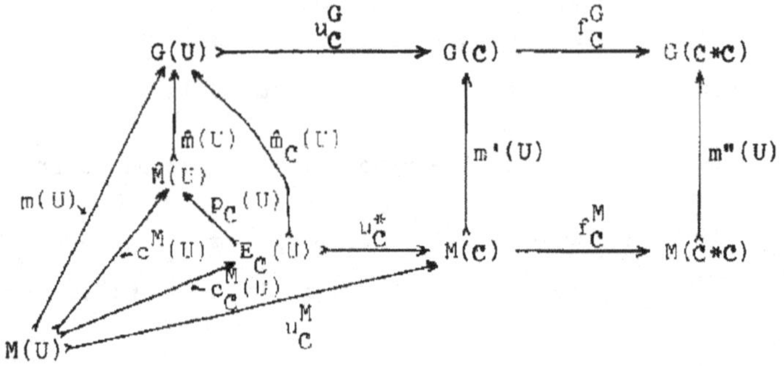

With this and the similar diagram for n , we obtain a diagram:

(19)

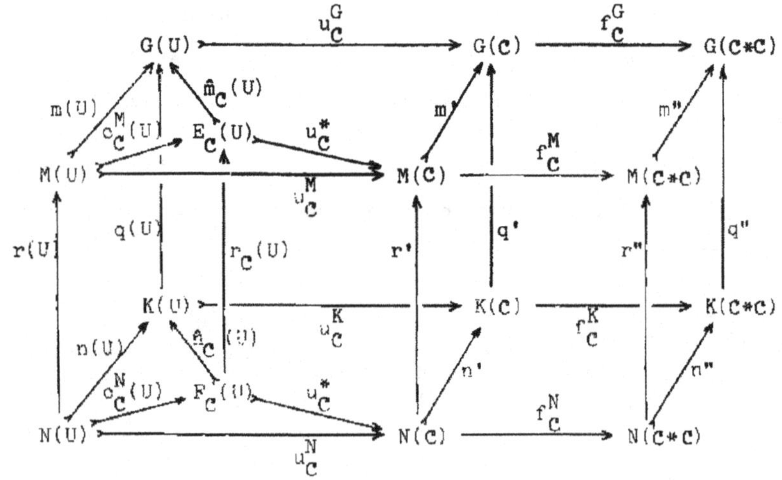

where F is used instead of E in the construction of $\hat{N}$ and $r_C(U)$ is induced on equalizers (considering the similar diagram with g's instead of f's); the diagram commutes, since all faces already commute except perhaps two of the squares containing $r_C(U)$, and these commute since u_C^G and u_C^* are monomorphisms. In addition, since products and equalizers preserve pullbacks, we see that $m'r' = q'n'$, $m''r'' = q''n''$ are pullbacks, and then $q(U)\ \hat{n}_C(U) = \hat{m}_C(U)\ r_C(U)$ is a pullback, too.

Therefore $\operatorname{Im} \hat{n}_C(U) = q(U)^s \operatorname{Im} \hat{m}_C(U)$. Now we assume that $\hat{m}$ is an isomorphism, and apply II.4.1. Since $\hat{m}(U)$ is induced by the by the $\hat{m}_C(U)$, and is a regular epimorphism, we have $\bigvee_{C\in\mathfrak{C}(U)} \operatorname{Im} \hat{m}_C(U) = 1$; by (C_3') it follows that $\bigvee_{C\in\mathfrak{C}(U)} \operatorname{Im} \hat{n}_C(U) = 1$, and the monomorphism $\hat{n}(U)$ induced by all $\hat{n}_C(U)$ is also a regular epimorphism, hence is an isomorphism, q.e.d.

With this lemma, it is easy to show that the pullback axiom holds in $\mathfrak{F}(X,G)$. Let $fg' = gf'$ be a pullback in $\mathfrak{F}$, where f is a regular epimorphism . Let (m,p) be a regular decomposition of f in $\mathfrak{P}$, and $mg_1' = gm'$, $pg_1'' = g_1'p'$ be pullbacks in $\mathfrak{P}$. Juxtaposition yields a pullback $(mp)g_1'' = g(m'p')$ which we may assume to be $fg' = gf'$; then $f' = m'p'$. Since $\mathfrak{P}$ is regular, (m',p') is a regular decomposition of f' in $\mathfrak{P}$. Now $\hat{m}$ is an isomorphism, by 3.6 ; by the lemma, $\hat{m}'$ is also an isomorphism, and then f' is a regular epimorphism in $\mathfrak{F}$. Thus we have proved that $\mathfrak{F} = \mathfrak{F}(X,G)$ is a regular category.

That it is a C_3 regular category is easily deduced from 3.7. First we have seen that it is cocomplete. Now let $\mathfrak{D}$ be a finite diagram in $[I,\mathfrak{F}]$, where I is a directed preordered set; we want to prove that $\varinjlim_{\mathfrak{F}}\lim_{\mathfrak{F}}\mathfrak{D} = \lim_{\mathfrak{F}}\varinjlim_{\mathfrak{F}}\mathfrak{D}$, where $\lim_{\mathfrak{F}}$ and $\varinjlim_{\mathfrak{F}}$ are limit and directed colimit functors, respectively, relative to $\mathfrak{F}$. We know that $\varinjlim_{\mathfrak{F}} = (\varinjlim_{\mathfrak{P}})^{\wedge}$, and that $\lim_{\mathfrak{F}}$ and $\lim_{\mathfrak{P}}$ are the same as long as we consider diagrams of sheaves only. Hence it follows from 3.7 that

$$\begin{aligned}\varinjlim_{\mathfrak{F}}\lim_{\mathfrak{F}}\mathfrak{D} &= \varinjlim_{\mathfrak{F}}\lim_{\mathfrak{P}}\mathfrak{D} = (\varinjlim_{\mathfrak{P}}\lim_{\mathfrak{P}}\mathfrak{D})^{\wedge} = (\lim_{\mathfrak{P}}\varinjlim_{\mathfrak{P}}\mathfrak{D})^{\wedge} = \\ &= \lim_{\mathfrak{F}}(\varinjlim_{\mathfrak{P}}\mathfrak{D})^{\wedge} = \lim_{\mathfrak{F}}\varinjlim_{\mathfrak{F}}\mathfrak{D} ,\end{aligned}$$

since $\mathfrak{P}$ is a C_3 [in fact C_4] regular category. The proof of the theorem is now complete.

3. The reader will no doubt have noticed that the proofs in this section make very little use of the techniques developed in part I. Of course this is more apparent than real since we did use them to establish theorem II.4.4 on which these proofs hang. All the same this

leaves a possibility that regularity is not needed for these results, if we start from a category in which the conclusion of theorem II.4.4 (in the non-C_1^* cases) holds. In view of I.1.6, this simply means (if we still assume cocompleteness or a minimum thereof) that we wish to go on without the pullback axiom.

In this case, we exclude the proof that the pullback axiom holds in $\mathfrak{F}(X,\mathcal{G})$; temporarily excluding the manipulation of subobjects at the end of the proof of 3.3, we see that all other proofs go through, if we assume that directed colimits are exact in $\mathcal{G}$. Now what we temporarily excluded is just a manipulation of subobjects and therefore should not require the pullback axiom. Indeed for this we need only assume that the equalizer property II.4.3 holds in $\mathcal{G}$ and that there are well-behaved images in $\mathcal{G}$; for this we can assume that $\mathcal{G}$ has regular decompositions (it suffices to assume that $\mathcal{G}$ has coequalizers), or even that $\mathcal{G}$ has strong decompositions (for then most of the results in section I.3 go through). In addition to this, and the property expressed by II.4.4 (in the non-C_1^* case), the assumptions on $\mathcal{G}$ are: completeness, exact directed colimits. In the results, the contents of 3.4, 3.5, 3.7 are (except for the regularity of $\mathfrak{F}(X,\mathcal{G})$) all saved. In other words, one still needs decompositions, but not the pullback axiom nor full cocompleteness.

The same remarks apply to the results of the next section, even though we formulate them for regular categories; however, Gray's condition $\mathfrak{F}_2$ is no longer a consequence of II.4.2 and must be added to the irregular hypotheses on $\mathcal{G}$ [there is no need to add $\mathfrak{F}_1$ as it can be seen that the conclusion of II.4.4 is stronger]; one also needs the full strength of II.4.4 (= including the C_1^* case).

4. STALK PROPERTIES.

1. We now assume that X is a topological space (not an arbitrary Grothendieck topology). We still let $\mathcal{G}$ be a C_4 regular category.

For each presheaf $P \in \mathcal{P}(X,\mathcal{G})$, the stalk P_x of P at $x \in X$ is $\varinjlim_{U \ni x} P(U)$. These can be used to define a stalk functor S of $\mathcal{P}(X,\mathcal{G})$ into the product category $\mathcal{G}^X$ (i.e. the functor category $[X,\mathcal{G}]$ in which X now denotes the obvious discrete category); namely, S sends P onto $(P_x)_{x \in X}$, and similarly for morphisms. Since $\mathcal{G}$ is, in particular, C_3, the stalk functor is exact.

We now observe that $\mathcal{G}$ satisfies all the axioms for an $\mathfrak{I}$-category as defined in [31] (see also [10],[11]) except for being locally small: in particular $\mathfrak{I}_1$ is part of the hypothesis and $\mathfrak{I}_2$ follows from II.4.2. Thus a great number, in fact most, of the results in [31] (in the non-abelian case) hold in our situation (the major exception being the existence of the associated sheaf, which we obtained previously). Specifically, we need to know that P and its associated sheaf have the same stalks (i.e. there is a natural isomorphism $S(P) \cong S(\hat{P})$); also, for each presheaf P, the presheaf $\bar{P}$ defined by: $\bar{P}(U) = \prod_{x \in U} P_x$, whose restriction maps are projections, is a sheaf, and a morphism of presheaves is defined by $m_P(U) = \bigtimes_{x \in U} P_{U,x} : P(U) \longrightarrow \bar{P}(U)$, and is a monomorphism if P is a monopresheaf. Then S is still exact on $\mathfrak{I}(X,\mathcal{G})$.

2. More can be proved if furthermore $\mathcal{G}$ is C_1^*. The basic result is the lemma which follows. We are indebted to VanOsdol for the remark that it shows $\mathfrak{I}(X,\mathcal{G})$ is cotripleable under $\mathcal{G}^X$ (more precisely, S_* is cotripleable).

Lemma 4.1. If $\mathcal{G}$ is C_1^*, then S reflects isomorphisms.

<u>Proof</u>. We have to show that if $f : F \longrightarrow G$ is a morphism of sheaves and $f_x : F_x \longrightarrow G_x$ is an isomorphism for all $x \in X$, then f is an isomorphism. From f we obtain a commutative diagram

$$(20) \qquad \begin{array}{ccc} \overline{F} & \xrightarrow{\ \overline{f}\ } & \overline{G} \\ {\scriptstyle m_F}\uparrow & & \uparrow{\scriptstyle m_G} \\ F & \xrightarrow[\ f\ ]{} & G \end{array}$$

(where $\overline{f}(U) = \prod_{x \in U} f_x$), where m_F, m_G are [pointwise] monomorphisms. On this diagram it is clear that, if $\overline{f}$ is a monomorphism (e.g. if $S(f)$ is an isomorphism) then f is also a monomorphism. Then the lemma follows at once from the more general fact that, when f is a monomorphism, (20) is a pullback. We now prove this property.

First G is C_1^* and so it follows from II.4.4 that an isomorphism

$$t_F(U) : \varinjlim_{\tau \in T} \prod_{x \in U} F(\tau x) \longrightarrow \prod_{x \in U} \varinjlim_{x \in V \subseteq U} F(V) = \overline{F}(U)$$

is induced by all $\qquad p_\tau^F = \prod_{x \in U} F_{\tau x, x} : \prod_{x \in U} F(\tau x) \longrightarrow \prod_{x \in U} F_x = \overline{F}(U)$

$(\tau \in T)$; there $T = \prod_{x \in U} \{V ; x \in V \subseteq U, V \text{ open}\}$, in other words T is the set of all mappings τ which to every $x \in U$ assign an open neighborhood $\tau x \subseteq U$ of x. From here on, we identify each $\tau \in T$ and the corresponding open covering $(\tau x)_{x \in U}$ of U. A similar description of $\overline{G}(U)$ can be given, in terms of morphisms p_τ^G (with the same T).

For each $\tau \in T$ we have a commutative diagram

$$(21) \qquad \begin{array}{ccccc} F(U) & \xrightarrow{\ u_\tau^F\ } & F(\tau) & \xrightarrow{\ f_\tau^F\ } & F(\tau * \tau) \\ {\scriptstyle f(U)}\downarrow & & \downarrow{\scriptstyle f'_\tau(U)} & & \downarrow{\scriptstyle f''_\tau(U)} \\ G(U) & \xrightarrow[\ u_\tau^G\ ]{} & G(\tau) & \xrightarrow[\ f_\tau^G\ ]{} & G(\tau * \tau) \end{array}$$

(where $f'_\tau(U) = \prod_{x \in U} f(\tau x)$ etc.); we now assume that f is a monomorphism, so that we have monomorphisms in the diagram as indicated. There is a similar diagram, with f^F_τ, f^G_τ replaced by g^F_τ, g^G_τ. We now claim that the left square in this diagram is a pullback. Assume indeed that $f'a = u^G b$. Then $f''f^F a = f^G f' a = f^G u^G b = g^G u^G b = \dots = f'' g^F a$; since F is a sheaf, and f'' is a monomorphism, it follows that $a = u^F c$ for some c; then also $u^G b = f'a = f'u^F c = u^G f(U)c$ and $b = f(U)c$. The factorization is unique since, say, u^G is a monomorphism.

Taking directed colimits over T yields a pullback to which we attach the isomorphisms $t_F(U)$, $t_G(U)$ to obtain the diagram below:

$$
(22)\qquad
\begin{array}{ccccc}
F(U) & \longrightarrow & \varinjlim F(\tau) & \xrightarrow[\cong]{t_F(U)} & \bar{F}(U) \\
{\scriptstyle f(U)}\downarrow & \text{PB} & \downarrow & & \downarrow{\scriptstyle \bar{f}(U)} \\
G(U) & \longrightarrow & \varinjlim G(\tau) & \xrightarrow[\cong]{t_G(U)} & \bar{G}(U)
\end{array}
$$

The top row yields a morphism $F(U) \longrightarrow \bar{F}(U)$ which is the colimit of $p^F_\tau u^F_\tau$. Now

$$p^F_\tau u^F_\tau = (\prod_{x \in U} F_{\tau x, x})(\underset{x \in U}{X} F_{U, \tau x}) = \underset{x \in U}{X} F_{U,x} = m_F(U) .$$

Hence the composite morphism in the top row is just $m_F(U)$. The bottom row similarly yields $m_G(U)$. Hence if we forget the middle column in (22), the resulting pullback is but (20) evaluated at U. It follows that (20) is a pullback, which completes the proof of the lemma.

The obvious application of the lemma is as follows. Let $\mathcal{D}$ be a diagram of sheaves and $(a_i)_{i \in I}$ be a cocompatible family for $\mathcal{D}$ (in $\mathcal{F}(X,\mathcal{A})$). Assume that $((a_i)_x)_{i \in I}$ is a colimit of $\mathcal{D}_x$ (in $\mathcal{A}$) for every $x \in X$. If a is the morphism induced to the colimit by the cocompatible family $(a_i)_{i \in I}$ then since S preserves colimits a_x is an isomorphism for every x; hence a is an isomorphism, and $(a_i)_{i \in I}$ is in fact a colimit of $\mathcal{D}$. This is expressed by saying that "colimits can safely be computed on the stalks" [we borrowed the expression from Van Osdol]. The same argument applies to anything that is preserved by S, which includes finite limits, and regular decompositions. Thus:

Theorem 4.2. Let G be a C_4, C_1^* regular category. For any topological space X, all colimits, finite limits and regular decompositions in $\mathfrak{F}(X,G)$ can safely be computed on the stalks.

REFERENCES

[1] BARR, M.: Relational algebras. Reports of the Midwest Category Seminar IV, 39-55. Springer Lecture Notes 137 (1970).

[2] BARR, M.: Factorizations, generators and rank. (Preprint)

[3] BARR, M.: Non-abelian full embedding, I. (Preprint)

[4] BARR, M.: Non-abelian full embedding, II. (Preprint) (For the last two references, see also the announcement in the Reports of the Midwest Category Seminar V, 205-208.)

[5] BENABOU, J.: Introduction to bicategories. Reports of the Midwest Category Seminar I, 1-77. Springer Lecture Notes 47 (1967).

[6] BUCUR and DELEANU: Introduction to the theory of categories and functors. John Wiley and Sons, 1968.

[7] COHN, P.M.: Universal Algebra. Harper and Row, 1965.

[8] ECKMANN, B. [editor]: Seminar on triples and categorical Homology theory. Springer Lecture Notes 80 (1969).

[9] FOLKS, The: Folk theorems.(Unpublished)(we hope)

[10] GRAY, J.S.: Sheaves with values in a category. Notes, Columbia University, 1962.

[11] GRAY, J.S.: Sheaves with values in a category. Topology 3 (1965) 1-18.

[12] GRAY, J.S.: Review of [16], MR 26 (1963) #1887.

[13] GRILLET, P.A.: Morphismes spéciaux et décompositions. C. R. Acad. Sci. Paris 266 (1968) sér.A, 397-398; Quelques propriétés des catégories non-abéliennes, ibid. 550-552; La suite exacte d'homologie dans une catégorie non-abélienne, ibid. 604-606.

[14] GRILLET, P.A.: Directed colimits and sheaves in some non-abelian categories. Reports of the Midwest Category Seminar V, 36-69. Springer Lecture Notes 195 (1971).

[15] GROTHENDIECK, A.: Sur quelques points d'Algèbre homologique. Tohoku Math. J. 9 (1957) 119-221.

[16] HELLER, A. and ROWE, K.A.: On the category of sheaves. Amer. J. Math. 84 (1962) 205-216.

[17] HERRLICH, H.: Topologische reflexionen und Coreflexionen. Springer Lecture Notes 78 (1968).

[18] HILTON, P.: Catégories non-abéliennes. Notes, Université de Montréal (1964).

[19] ISBELL, J.R.: Some remarks concerning categories and subspaces. Canad. J. Math. 9 (1957) 563-577.

[20] ISBELL, J.R.: Subobjects, adequacy, completeness and categories of algebras. Rozprawy Mat. 36 (1964).

[21] ISBELL, J.R.: Structure of categories. Bull. Amer. Math. Soc. 72 (1966) 619-655.

[22] ISBELL, J.R. and HOWIE, J.M.: Epimorphisms and dominions, II. J. Algebra 6 (1967) 7-21.

[23] KELLY, G.M.: Monomorphisms, epimorphisms and pullbacks. J. Austral. Math. Soc. 9 (1969) 124-142.

[24] KENNISON, J.F.: Full reflective subcategories and generalized coverings. Ill. J. Math. 12 (1968) 353-365.

[25] LAWVERE, F.V.: Functorial semantics of algebraic theories. Doct. Diss., Columbia University (1963).

[26] MAC LANE, S.: Groups, categories and duality. Proc. Nat. Acad. Sci. U.S.A. 34 (1948) 263-267.

[27] MAC LANE, S.: Duality for groups. Bull. Amer. Math. Soc. 56 (1950) 485-516.

[28] MAC LANE, S.: An algebra of additive relations. Proc. Nat. Acad. Sci. U.S.A. 47 (1961) 1043-1051.

[29] MAC LANE, S.: Homology. Springer, 1963.

[30] MANES, E.G.: A triple miscellany: some aspects of the theory of algebras over a triple. Doct. Diss., Wesleyan University (1967)

[31] MITCHELL, B.: Theory of categories. Academic Press, 1965.

[32] PIERCE, R.S.: Introduction to the theory of abstract algebras. Holt, Rinehart and Winston, 1968.

[33] PUPPE, D.: Korrespondenzen in Abelschen Kategorien. Math. Ann. 148 (1962) 1-30.

[34] SEMADENI, Z.: Projectivity, injectivity and duality. Rozprawy Mat. 35 (1963).

[35] VAN OSDOL, D.H.: Sheaves of algebras (to appear).

SHEAVES IN REGULAR CATEGORIES

Donovan H. Van Osdol

INTRODUCTION

The investigation presented here was inspired by the following conjecture of Michael Barr. Let X be a topological space, $\underline{P}(X,\underline{A})$ the category of presheaves on X with values in the category $\underline{A}$, and T: $\underline{P}(X,\underline{A}) \to \underline{P}(X,\underline{A})$ the Godement standard construction [4] with unit ηP: $P \to TP$ for any presheaf P. Then the sheaf associated to P is the equalizer of ηTP and $T\eta P$. I will prove this (when $\underline{A}$ is a regular category with some extra conditions) as a corollary of the stronger result: the "stalk functor" restricted to sheaves on X with values in $\underline{A}$ is cotripleable [3]. This result also yields a new description of the category $\underline{F}(X,\underline{A})$ of sheaves. Propositions III.4 and III.5 have been derived by Grillet, using completely different techniques.

Throughout this paper, if A and B are objects of a category $\underline{A}$ then (A,B) will denote the set of $\underline{A}$-morphisms from A to B. If $\underline{A}$ is a small category and $\underline{B}$ is a category then $\underline{B}^{\underline{A}}$ will denote the category whose objects are covariant functors $\underline{A} \to \underline{B}$ and whose morphisms are natural transformations of functors. A symbol X will denote either the object X or the identity morphism on X. Any other undefined symbols or terms are either standard, or else can be found in the papers of Barr and Grillet in this volume.

I. Transfer Theorems for Triples

Our starting point is the transfer theorem [5], [6]: If $\underline{A}$ is a regular category, $\mathbb{T}$ a triple on $\underline{A}$ which preserves regular epimorphisms, then $\underline{A}^{\mathbb{T}}$ is a regular category, and $U^{\mathbb{T}}$: $\underline{A}^{\mathbb{T}} \to \underline{A}$ preserves decompositions. When $\underline{A}$ is assumed to satisfy additional conditions, it is natural to ask whether these new conditions are also transferred to $\underline{A}^{\mathbb{T}}$. The extra conditions in which we are interested are those which make $\underline{A}^{\mathbb{T}}$ a "good category for sheaf theory".

Throughout this section, $\underline{B}$ will be a cocomplete category, and $\underline{A}$ will be a regular category. The triple $\mathbb{T}$ on $\underline{A}$ will be induced by an adjoint pair of functors $U \dashv F$ in which $U: \underline{B} \to \underline{A}$ is tripleable and commutes with directed colimits. Moreover, $T = UF$ will be assumed to preserve regular epimorphisms. Notation will be as in [5].

<u>Proposition I.1</u>. If $\underline{A}$ satisfies Grillet's property C_3', then $\underline{B}$ also satisfies C_3'.

<u>Proof</u>: Let $f: B' \to B$ be a morphism in $\underline{B}$, and let $\{B_\alpha | \alpha \in I\}$ be a directed family of subobjects of B. Form the following pullback diagrams in $\underline{B}$:

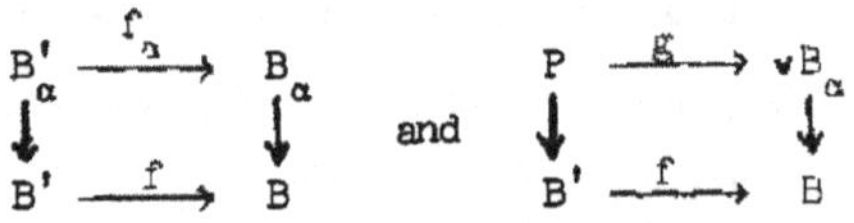

Our task is to show that $\vee B_\alpha'$ is isomorphic to P. Grillet has shown that $\vee B_\alpha \cong \operatorname{colim} B_\alpha$, and thus by our assumption on U, $U(\vee B_\alpha) \cong \vee UB_\alpha$. Let $h: \vee B_\alpha' \to P$ be the unique mapping induced by the pullback condition:

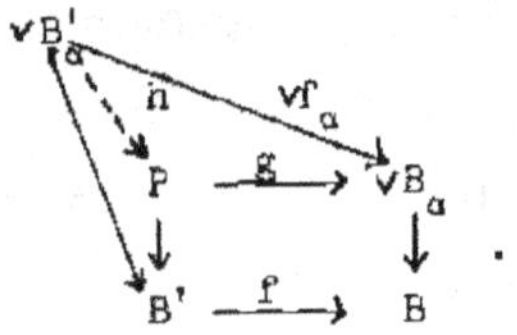

Applying U to this diagram yields:

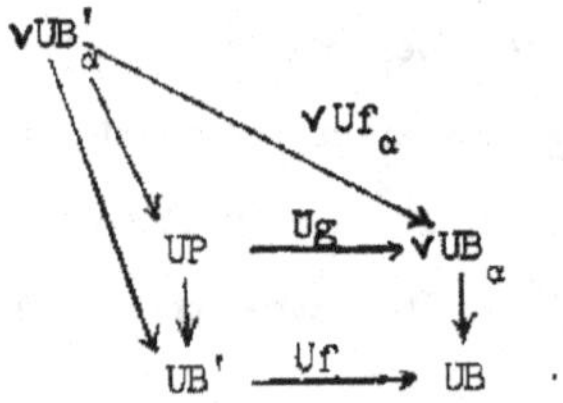

Now the outer square is a pullback because $\underline{A}$ satisfies C_3', and the inner square is a pullback because U preserves limits. Thus Uh is an isomorphism. But U reflects isomorphisms (by Beck's theorem [3], [6]) and hence h is an isomorphism.

<u>Proposition I.2</u>. If $\underline{A}$ satisfies Grillet's condition C_3, then $\underline{B}$ also satisfies C_3.

<u>Proof</u>: Given a direct system $\{B_\alpha; f_{\alpha\beta} | \alpha, \beta \in I\}$ in $\underline{B}$ such that each $f_{\alpha\beta}$ is a monomorphsim, we have the map $f_\alpha = B_\alpha \to \amalg B_\beta \to \operatorname{colim} B_\beta$. We are to show that $B_\alpha \to \amalg B_\beta$ is a monomorphsim for each $\alpha \in I$, for which it suffices to show that f_α is monic. But we have

$$Uf_\alpha = (UB_\alpha \to U \operatorname{colim} B_\beta) \approx (UB_\alpha \to \operatorname{colim} UB_\beta) ,$$

which Grillet has shown to be monic. Now U is faithful and thus f_α is a monomorphism. Turning now to condition C_3'', let $\{K_\alpha \overset{x_\alpha}{\underset{y_\alpha}{\rightrightarrows}} B \xrightarrow{p_\alpha} B_\alpha\}$ be a directed family of congruences on B in $\underline{B}$. Then $UK_\alpha \rightrightarrows UB \to UB_\alpha$ is a directed family of congruences on UB, so that by Grillet's work we know $\vee UK_\alpha \rightrightarrows UB \longrightarrow \operatorname{colim} UB_\alpha$ is a congruence in $\underline{A}$. Let $p = \operatorname{colim} p_\alpha$. Since

$$U(\vee K_\alpha) \approx \vee UK_\alpha \rightrightarrows UB \xrightarrow{Up} U \operatorname{colim} B_\alpha$$

is an equalizer in $\underline{A}$ and U creates limits [6], it follows that $\vee K_\alpha \rightrightarrows B \xrightarrow{p} \operatorname{colim} B_\alpha$ is an equalizer in B. Hence $\vee K_\alpha \rightrightarrows B$ is a congruence. Together with I.1 we now have demonstrated that $\underline{B}$ is a C_3 category.

<u>Proposition I.3</u>. If $\underline{A}$ satisfies Grillet's condition C_4, then $\underline{B}$ also satisfies C_4.

<u>Proof</u>: Since U creates limits [6], $\underline{B}$ is complete. Proposition I.2 says that $\underline{B}$ satisfies C_3, so it remains to verify F_1. Let $\{\{B_i | i \in I_\lambda\} | \lambda \in \Lambda\}$ be a non-empty family of non-empty directed families of subobjects of B in $\underline{B}$, with the I_λ pairwise disjoint. Let $S = \{\tau\colon \Lambda \longrightarrow \cup I_\lambda | \tau(\lambda) \in I_\lambda \text{ for all } \lambda \in \Lambda\}$. We want to show that the natural mapping $f\colon \bigvee_{\tau \in S} \bigwedge_{\lambda \in \Lambda} B_{\tau(\lambda)} \longrightarrow \bigwedge_{\lambda \in \Lambda} \bigvee_{i \in I_\lambda} B_i$ is an isomorphism. Applying U and using the directedness of S, we get:

$$\begin{array}{ccc} U[\vee\wedge B_{\tau(\lambda)}] & \xrightarrow{Uf} & U[\wedge\vee B_1] \\ \wr\| & & \wr\| \\ \vee\wedge UB_{\tau(\lambda)} & \xrightarrow{\cong} & \wedge\vee UB_1 \end{array} .$$

Hence Uf is an isomorphism, and so is f .

Proposition I.4. If in $\underline{A}$ the product of any family of regular epimorphisms is itself regular epi, then the same is true in $\underline{B}$.

Proof: It is shown in the transfer theorem that a morphism in $\underline{B}$ is regular epi if and only if U of it is regular epi in $\underline{A}$. Let $\{f_\alpha : B'_\alpha \to B_\alpha | \alpha \in I\}$ be a family of regular epis in $\underline{B}$. Then $U(\Pi f_\alpha) = \Pi U f_\alpha$ is a product of regular epis in $\underline{A}$, hence is regular epi, and therefore Πf_α is regular epi.

Proposition I.5. If directed colimits commute with finite limits in $\underline{A}$, then the same is true in $\underline{B}$.

Proof: The functor U commutes with directed colimits, (finite) limits, and it reflects isomorphisms.

Proposition I.6. Suppose the following condition holds in the category $\underline{A}$: for each non-empty family $\{X^\lambda | \lambda \in \Lambda\}$ of direct systems over pairwise disjoint directed sets I_λ there is a natural isomorphism $\prod_{\lambda\in\Lambda} \operatorname{colim} X^\lambda \cong \operatorname{colim}_{\tau\in S} \prod_{\lambda\in\Lambda} A_{\tau(\lambda)}$ where $S = \{\tau : \Lambda \longrightarrow \bigcup_{\lambda\in\Lambda} I_\lambda | \tau(\lambda) \in I_\lambda \text{ for all } \lambda\}$. Then the same condition holds in $\underline{B}$.

Proof: Let $\{X^\lambda\}$ be such a family in $\underline{B}$. We have the natural map

$f: \operatorname{colim} \Pi B_{\tau(\lambda)} \to \Pi \operatorname{colim} X^\lambda$, and

$$\begin{array}{ccc} U(\operatorname{colim} \Pi B_{\tau(\lambda)}) & \xrightarrow{Uf} & U(\Pi \operatorname{colim} X^\lambda) \\ \wr\| & & \wr\| \\ \operatorname{colim} \Pi UB_{\tau(\lambda)} & \xrightarrow{\cong} & \Pi \operatorname{colim} UX^\lambda \end{array} .$$

But U reflects isomorphisms, so f is an isomorphism.

II. Transfer Theorems for Cotriples

We give here those, and only those, theorems which will be directly useful in our study of sheaves. At least one of the results is known, but we know of no reference for it.

Proposition II.1. If $Q: \underline{A} \to \underline{B}$ has a left adjoint then Q preserves monomorphisms.

Proof: See [7].

Proposition II.2. Suppose $\underline{A}$ and $\underline{B}$ have kernel pairs, and $S: \underline{B} \longrightarrow \underline{A}$ preserves kernel pairs. Then S preserves monomorphisms.

Proof: Recall that $f: B \longrightarrow B'$ is a monomorphism if and only if in the kernel pair diagram, $K \underset{k_2}{\overset{k_1}{\rightrightarrows}} B \xrightarrow{f} B'$, we have $k_1 = k_2$. Thus if $f: B \to B'$ is a monomorphism in $\underline{B}$ then $K \underset{k}{\overset{k}{\rightrightarrows}} B \xrightarrow{f} B'$ is a kernel pair implies $SK \underset{Sk}{\overset{Sk}{\rightrightarrows}} SB \xrightarrow{Sf} SB'$ is a kernel pair in $\underline{A}$. Thus Sf is a monomorphism.

Proposition II.3. If $S: \underline{B} \to \underline{A}$ is cotripleable then S creates colimits. This means that if $\Gamma: \underline{D} \to \underline{B}$ is a small diagram and $S\Gamma: \underline{D} \to \underline{A}$ has a colimit in $\underline{A}$ then Γ has a colimit in $\underline{B}$ and S preserves it.

Proof: See [6] for a proof of the dual assertion.

Theorem II.4. Suppose that $\underline{A}$ has finite limits and that the cotriple $\mathbf{G} = (G, \varepsilon, \delta)$ on $\underline{A}$ commutes with finite limits. Then $S_{\mathbf{G}}: \underline{A}_{\mathbf{G}} \to \underline{A}$ creates finite limits.

Proof: It suffices to prove that $S_{\mathbf{G}}$ creates finite products and equalizers. We first deal with products. Let (A_1, β_1), (A_2, β_2) be in $|\underline{A}_{\mathbf{G}}|$ and let β be the unique morphism making the following diagram commute for $j = 1, 2$:

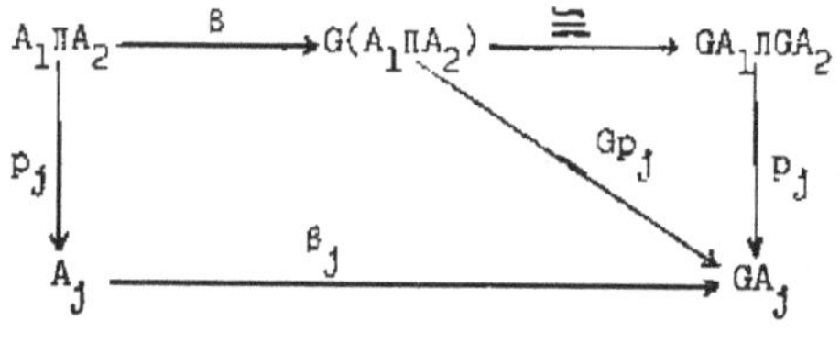

We claim that $(A_1 \Pi A_2, \beta)$ is in $|\underline{A}_{\mathbf{G}}|$.

i) $\varepsilon(A_1 \Pi A_2) \cdot \beta = (\varepsilon A_1 \Pi \varepsilon A_2) \cdot \beta = (\varepsilon A_1 \Pi \varepsilon A_2) \cdot (\beta_1 \cdot p_1 \Pi \beta_2 \cdot p_2)$

$= \varepsilon A_1 \cdot \beta_1 \cdot p_1 \Pi \varepsilon A_2 \cdot \beta_2 \cdot p_2 = p_1 \Pi p_2 = A_1 \Pi A_2$.

ii) $G\beta \cdot \beta = G(\beta_1 \cdot p_1 \Pi \beta_2 \cdot p_2) \cdot (\beta_1 \cdot p_1 \Pi \beta_2 \cdot p_2)$

$= (G(\beta_1 \cdot p_1) \Pi\, G(\beta_2 \cdot p_2)) \cdot (\beta_1 \cdot p_1 \Pi \beta_2 \cdot p_2)$

$= G\beta_1 \cdot \beta_1 \cdot p_1 \Pi\, G\beta_2 \cdot \beta_2 \cdot p_2$

$= \delta A_1 \cdot \beta_1 \cdot p_1 \Pi\, \delta A_2 \cdot \beta_2 \cdot p_2$

$= (\delta A_1 \cdot p_1 \Pi\, \delta A_2 \cdot p_2) \cdot (\beta_1 \cdot p_1 \Pi \beta_2 \cdot p_2)$

$= \delta(A_1 \Pi A_2) \cdot \beta$.

Moreover, given $f_j\colon (A,\alpha) \to (A_j,\beta_j)$ for $j = 1, 2$ in $\underline{A}_{\mathfrak{G}}$, there is a unique $f_1 \times f_2\colon A \to A_1 \Pi A_2$ in $\underline{A}$ such that $p_j \cdot (f_1 \times f_2) = f_j$ for $j = 1, 2$. We need only show that $f_1 \times f_2$ is in $\underline{A}_{\mathfrak{G}}$. But in the diagram:

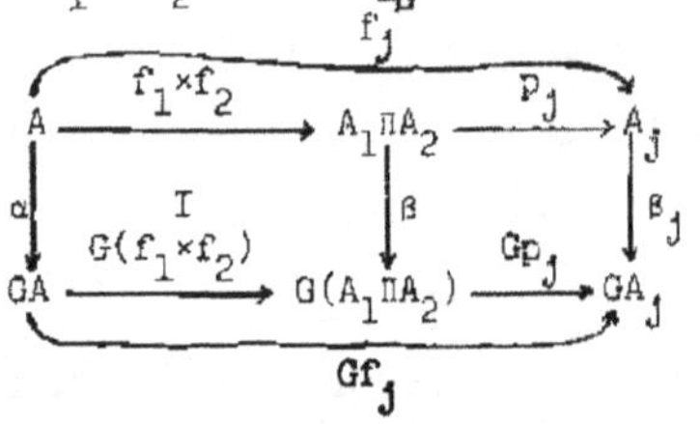

everything commutes, except possibly square I , for $j = 1, 2$. Thus by uniqueness of maps into a product, square I also commutes, and $f_1 \times f_2$ is a $\mathfrak{G}$-homomorphism. It follows that $(A_1,\beta_1) \Pi (A_2,\beta_2) = (A_1 \Pi A_2, \beta)$, and $S_{\mathfrak{G}}$ preserves this product.

Turning now to equalizers, let $(A_1,\beta_1) \underset{g}{\overset{f}{\rightrightarrows}} (A_2,\beta_2)$ be a diagram in $\underline{A}_{\mathfrak{G}}$ and let $A \xrightarrow{e} A_1 \underset{g}{\overset{f}{\rightrightarrows}} A_2$ be the equalizer in $\underline{A}$. Since G preserves limits, there exists a unique $\alpha\colon A \to GA$ such that $Ge \cdot \alpha = \beta_1 \cdot e$. We claim that (A,α) is in $|\underline{A}_{\mathfrak{G}}|$.

i) $\beta_1 \cdot e \cdot \varepsilon A \cdot \alpha = \beta_1 \cdot \varepsilon A_1 \cdot Ge \cdot \alpha = Ge \cdot \alpha = \beta_1 \cdot e$

and since $\beta_1 \cdot e$ is monic, $\varepsilon A \cdot \alpha = A$.

ii) $G^2 e \cdot G\alpha \cdot \alpha = G\beta_1 \cdot Ge \cdot \alpha = G\beta_1 \cdot \beta_1 \cdot e$

$= \delta A_1 \cdot \beta_1 \cdot e = \delta A_1 \cdot Ge \cdot \alpha = G^2 e \cdot \delta A \cdot \alpha$

and since $G^2 e$ is monic (Proposition II.2),

$G\alpha \cdot \alpha = \delta A \cdot \alpha$.

Moreover, given $h: (C,\gamma) \to (A_1,\beta_1)$ in $\underline{A}_\oplus$ such that $f \cdot h = g \cdot h$, there exists a unique map $\bar{h}: C \to A$ in $\underline{A}$ such that $e \cdot \bar{h} = h$. We need only show that $\bar{h}$ is in $\underline{A}_\oplus$. But $Ge \cdot \alpha \cdot \bar{h} = \beta_1 \cdot e \cdot \bar{h} = \beta_1 \cdot h = Gh \cdot \gamma = Ge \cdot G\bar{h} \cdot \gamma$, and Ge monic yields $\alpha \cdot \bar{h} = G\bar{h} \cdot \gamma$. Hence $(A,\alpha) \xrightarrow{\;e\;} (A_1,\beta_1) \underset{g}{\overset{f}{\rightrightarrows}} (A_2,\beta_2)$ is the equalizer in $\underline{A}_\oplus$, and $S_\oplus$ preserves it.

<u>Theorem II.5</u>. Suppose that $\underline{A}$ is a regular category, that $S: \underline{B} \to \underline{A}$ is cotripleable, and that S commutes with finite limits. Then $\underline{B}$ is a regular category.

<u>Proof</u>: Because of the theorems already proved, and Tierney's theorem [1], it remains to prove that $\underline{B}$ satisfies the pullback condition for regular epis. Let $f: B_1 \to B_3$ be a regular epimorphism in $\underline{B}$ and form the pullback over any morphism $B_2 \to B_3$:

$$
\begin{array}{ccc}
X & \xrightarrow{\;f'\;} & B_2 \\
\downarrow & & \downarrow \\
B_1 & \xrightarrow{\;f\;} & B_3
\end{array}
$$

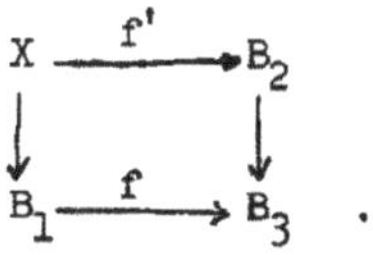

Now f being regular epi means it is the coequalizer of its kernel pair, and since S preserves finite limits and colimits, Sf is regular epi. Moreover

$$
\begin{array}{ccc}
SX & \xrightarrow{\;Sf'\;} & SB_2 \\
\downarrow & & \downarrow \\
SB_1 & \xrightarrow{\;Sf\;} & SB_3
\end{array}
$$

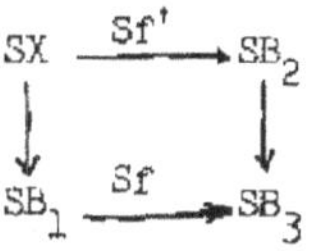

is a pullback in $\underline{A}$ so that Sf' is regular epi, that is, coequalizer of its kernel pair. But S creates finite limits and colimits, so that f' is the coequalizer of its kernel pair. Thus $\underline{B}$ is a regular category.

<u>Theorem II.6</u>. Assume that in the diagram:

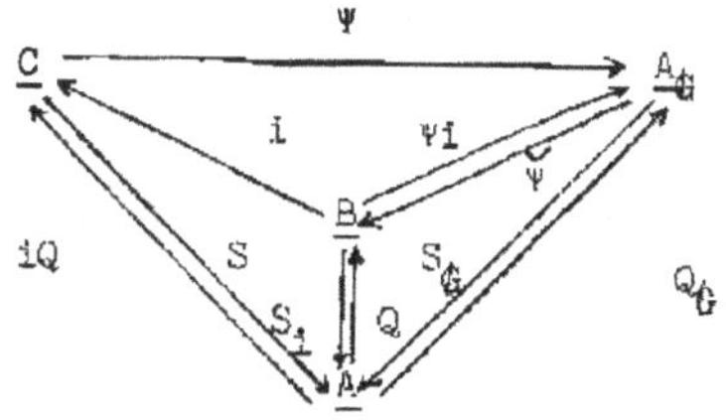

i is a full inclusion, $S \dashv iQ$ with associated cotriple $\mathbf{G}$. Then $Si \dashv Q$. Moreover, assuming that Si is cotripleable, $\check{\Psi}\Psi \dashv i$.

Proof: Given $A \in |\underline{A}|$, $B \in |\underline{B}|$ we have $(SiB,A) \approx (iB,iQA) = (B,QA)$. Also, given $C \in |\underline{C}|$, $(\check{\Psi}\Psi C,B) \approx (\check{\Psi}\Psi C,\check{\Psi}\Psi iB) \approx (\Psi i\check{\Psi}\Psi C,\Psi iB) \approx (\Psi C,\Psi iB) \approx (C,i\check{\Psi}\Psi iB) \approx (C,iB)$.

Proposition II.7. In the same situation as in Theorem II.6, assume that S preserves finite limits. A map $f\colon B_1 \to B_2$ in $\underline{B}$ is a regular epimorphism if and only if whenever $i(f) = m \cdot p$ is a mono-regular epi factorization, $\check{\Psi}\Psi(m)$ is an isomorphism.

Proof: Suppose that f is regular epi. We have:

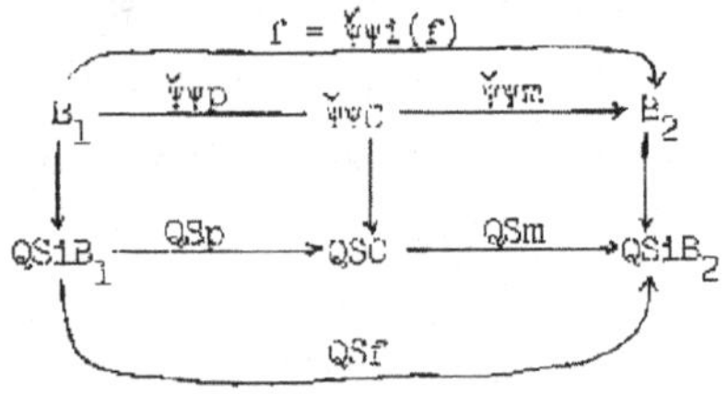

where the vertical maps are equalizers. Now since QSm is a monomorphism, so is $\check{\Psi}\Psi m$. On the other hand, f is regular epi and hence so is $\check{\Psi}\Psi m$. Thus $\check{\Psi}\Psi m$ is an isomorphism. Conversely, suppose that $\check{\Psi}\Psi m$ is an isomorphism. Now p is the coequalizer of $i(f)$'s kernel pair, and thus $\check{\Psi}\Psi p$ is the coequalizer of the kernel pair of f (recall how to compute colimits in reflective subcategories). But $\check{\Psi}\Psi m \cdot \check{\Psi}\Psi p = f$ and $\check{\Psi}\Psi m$ is an isomorphism. Thus $f \approx \check{\Psi}\Psi p$ is a regular epimorphism.

III. Sheaves

Let A be a complete, cocomplete, regular category in which the following condition holds: (*) Let $\{X^\lambda \mid \lambda \in \Lambda\}$ be a non-empty family of direct systems over pairwise disjoint directed preordered sets I_λ. Then there is a natural isomorphism:

$$\prod_{\lambda\in\Lambda} \operatorname{colim} X^\lambda \longrightarrow \operatorname*{colim}_{\tau\in S} \prod_{\lambda\in\Lambda} X_{\tau(\lambda)}$$

where $S' = \{\tau\colon \Lambda \longrightarrow \bigcup I_\lambda \mid \tau(\lambda) \in I_\lambda\}$. We also assume that directed colimits in $\underline{A}$ commute with finite limits.

Given a topological space X let $\underline{P}(X,\underline{A})$ (respectively $\underline{F}(X,\underline{A})$) be the category of presheaves (respectively sheaves) on X with values in $\underline{A}$. Let $|X|$ be the discrete category on the underlying set of X . Define $S\colon \underline{P}(X,\underline{A}) \to \underline{A}^{|X|}$ to be the stalk functor, that is, $(SP)x = P_x = \operatorname{colim} P(V)$ where the colimit is taken over the directed set consisting of open sets V which contain x . The definition of S on morphisms is obvious. Define $Q\colon \underline{A}^{|X|} \to \underline{F}(X,\underline{A})$ by $Q\{A_x\}V = \prod_{x\in V} A_x$ for each open set V in X , and similarly for morphisms. Clearly $Q\{A_x\}$ is a presheaf, and is in fact a sheaf [7]. Notice that QS is the Godement standard construction [4]. Let $i\colon \underline{F}(X,\underline{A}) \to \underline{P}(X,\underline{A})$ be the inclusion functor.

Proposition III.1. The functor S is left adjoint to iQ .

Proof: The adjunction morphisms $\eta\colon \underline{P}(X,\underline{A}) \to iQS$ and $\varepsilon\colon SiQ \to \underline{A}^{|X|}$ are defined in the following diagrams, where the notation is obvious.

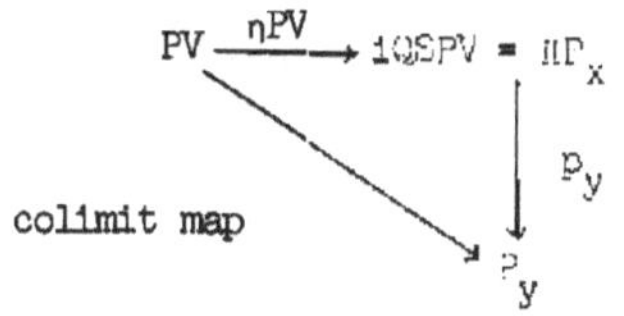

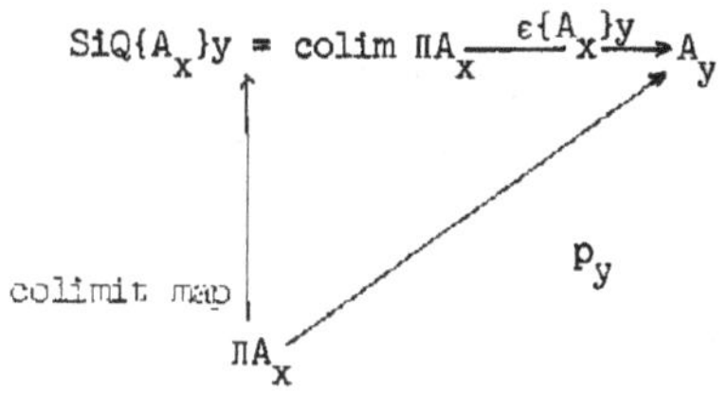

Theorem III.2. Let V be an open subset of X , $T = \{\tau\colon V \to \{\text{open subsets of } V\} \mid x \in \tau(x) \text{ for all } x \in V\}$ with induced preorder relation. For a presheaf P and $\tau \in T$ let $\times P_{V,\tau(x)}\colon PV \to \prod_{x\in V} P(\tau(x))$ be the obvious map. Then ηPV is the composition

$$PV \xrightarrow{\operatorname{colim}_{\tau\in T} \times P_{V,\tau(x)}} \operatorname{colim}_{\tau\in T} \prod_{x\in V} P(\tau(x)) \xrightarrow{\cong} \prod_{x\in V} \operatorname{colim}_{x\in W\subseteq V} P(W) .$$

Moreover, P is a monopresheaf if and only if ηP is a monomorphism.

Proof: The first assertion follows because each part of the diagram below commutes (recall condition (*) above).

$$\begin{array}{ccccccc}
PV & \xrightarrow{\times P_{V,\tau(x)}} & \prod_{x\in V} P(\tau(x)) & \xrightarrow{\text{colim map}} & \operatorname{colim}_{\tau\in T} \prod_{x\in V} P(\tau(x)) & \xrightarrow{\cong} & \prod_{x\in V} \operatorname{colim}_{x\in W\subseteq V} P(W) \\
 & \searrow^{P_{V,\tau(y)}} & \downarrow p_y & & \downarrow \operatorname{colim}_{\tau\in T} p_y & & \downarrow p_y \\
 & & P(\tau(y)) & \xrightarrow{\text{colim map}} & \operatorname{colim}_{\tau\in T} P(\tau(y)) & \xrightarrow{\cong} & \operatorname{colim}_{y\in W\subseteq V} P(W)
\end{array}$$

Now if P is a monopresheaf, then $\times P_{V,\tau(x)}$ is a monomorphism for each $\tau \in T$. As in Proposition II.2, since directed colimits preserve kernel pairs, $\operatorname{colim} \times P_{V,\tau(x)}$ is a monomorphism. Hence ηP is a monomorphism. The converse follows since ηP is the top row of the above diagram.

Theorem III.3. The functor Si: $\underline{F}(X,\underline{A}) \to \underline{A}^{|X|}$ is crudely cotripleable.

Proof: By Theorem II.6, $Si \dashv Q$. Moreover, Si preserves equalizers because directed colimits commute with finite limits. It remains to show that Si reflects equalizers. So suppose that $F \xrightarrow{f} G \overset{g}{\underset{h}{\rightrightarrows}} H$ is a diagram of sheaves such that $SiF \longrightarrow SiG \rightrightarrows SiH$ is an equalizer in $\underline{A}^{|X|}$. Then also $QSiF \longrightarrow QSiG \rightrightarrows QSiH$ is an equalizer in $\underline{F}(X,\underline{A})$, and we have the following diagram:

$$\begin{array}{ccccc}
F & \xrightarrow{f} & G & \overset{g}{\underset{h}{\rightrightarrows}} & H \\
\downarrow \eta F & & \downarrow \eta G & & \downarrow \eta H \\
QSiF & \xrightarrow{QSf} & QSiG & \overset{QSg}{\underset{QSh}{\rightrightarrows}} & QSiH
\end{array}$$

If the left-hand square is a pullback then we are done, for suppose K is a sheaf and k: $K \to G$ with $g \cdot k = h \cdot k$. Then $QSig \cdot \eta G \cdot k = \eta H \cdot g \cdot k = \eta H \cdot h \cdot k = QSih \cdot \eta G \cdot k$, so there exists a unique $\bar{k}$: $K \to QSiF$ such that $QSif \cdot \bar{k} = \eta G \cdot k$. Now by the pullback condition there exists a unique ℓ: $K \to F$ such that $\eta F \cdot \ell = \bar{k}$ and $f \cdot \ell = k$. If also $f \cdot \ell' = k$ then $QSif \cdot \eta F \cdot \ell' = QSif \cdot QSi\ell' \cdot \eta H$ $= QSik \cdot \eta K = \eta G \cdot k = QSif \cdot \bar{k}$, and $QSif$ monic implies $\eta F \cdot \ell' = \bar{k} = \eta F \cdot \ell$.

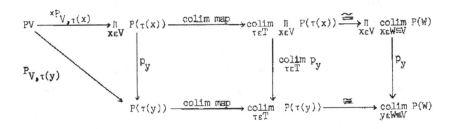

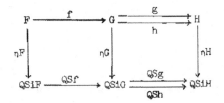

By Theorem III.2 ηF is monic, so $\ell' = \ell$ and f is the equalizer of g and h. To show that the above square is a pullback, note first that f is monic because $\eta G \cdot f = QSif \cdot \eta F$ is a monomorphism. Let V be an open subset of X and let $T = \{\tau\colon\ V \to \{\text{open subsets of } V\} \mid x \in \tau(x) \text{ for all } x \in V\}$. In the diagram:

$$\begin{array}{ccccc}
FV & \xrightarrow{\kappa F_{V,\tau(x)}} & \prod\limits_{x\in V} F(\tau(x)) & \rightrightarrows & \prod\limits_{x,y\in V} F(\tau(x)\wedge\tau(y)) \\
{\scriptstyle fV}\downarrow & & \downarrow{\scriptstyle \Pi f(\tau(x))} & & \downarrow{\scriptstyle \Pi f(\tau(x)\wedge\tau(y))} \\
GV & \xrightarrow{\kappa G_{V,\tau(x)}} & \prod\limits_{x\in V} G(\tau(x)) & \rightrightarrows & \prod\limits_{x,y\in V} G(\tau(x)\wedge\tau(y))
\end{array}$$

the rows are equalizers and the columns are monomorphisms (τ fixed in T). Hence the left-hand square here is a pullback. Since directed colimits commute with pull-backs, and condition (*) holds in $\underline{A}$ we see that

$$\begin{array}{ccccc}
 & & \xrightarrow{\qquad\qquad\eta FV\qquad\qquad} & & \\
FV & \longrightarrow & \operatorname*{colim}\limits_{\tau\in T} \prod\limits_{x\in V} F(\tau(x)) & \xrightarrow{\cong} & QSiFV \\
{\scriptstyle fV}\downarrow & & \downarrow{\scriptstyle \operatorname{colim} \Pi f\tau(x)} & & \downarrow{\scriptstyle QSfV} \\
GV & \longrightarrow & \operatorname*{colim}\limits_{\tau\in T} \prod\limits_{x\in V} G(\tau(x)) & \xrightarrow{\cong} & QSiGV \\
 & & \xrightarrow{\qquad\qquad\eta GV\qquad\qquad} & &
\end{array}$$

is a pullback. Hence Si reflects equalizers and is crudely cotripleable.

<u>Proposition III.4</u>. The functor $Si\colon\ \underline{F}(X,\underline{A}) \to \underline{A}^{|X|}$ creates colimits and finite limits, and $\underline{F}(X,\underline{A})$ is a regular cateogry. A map $f\colon\ P_1 \to P_2$ of sheaves is a regular epimorphism if and only if whenever $i(f) = m \cdot p$ is a mono-regular epi factorization in $\underline{P}(X,\underline{A})$, the associated sheaf map $\check{\Psi}\Psi m$ is an isomorphism.

<u>Proof</u>: This is a corollary of Theorem III.3 and the results in section II.

<u>Proposition III.5</u>. The reflection $\underline{P}(X,\underline{A}) \to \underline{F}(X,\underline{A})$ preserves mono-regular epi

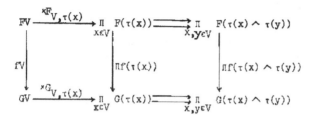

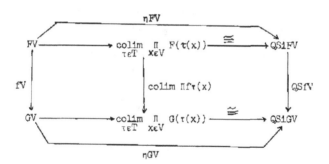

factorizations, colimits, and finite limits.

Proof: Since $S = Si\check{\Psi}\Psi$ preserves all of these and Si reflects them, $\check{\Psi}\Psi$ preserves them.

Theorem III.6. Suppose $U\colon \underline{B} \to \underline{A}$ is tripleable, $\underline{B}$ is cocomplete, $T = UF$ preserves regular epis, and U commutes with directed colimits. Then $S'i\colon \underline{F}(X,\underline{B}) \to \underline{B}^{|X|}$ is cotripleable and $U'\colon \underline{F}(X,\underline{B}) \to \underline{F}(X,\underline{A})$ is tripleable.

Proof: The hypotheses and theorems in section I combine to show that $\underline{B}$ is a complete, cocomplete, regular category in which directed colimits commute with finite limits and in which condition (*) holds. Hence by Theorem III.3, the stalk functor $S'i\colon \underline{F}(X,\underline{B}) \to \underline{B}^{|X|}$ is crudely cotripleable. The "underlying" functor $U'\colon \underline{F}(X,\underline{B}) \to \underline{F}(X,\underline{A})$ is defined by $(U'P)V = U(P(V))$ for each open subset V of X. Since U preserves equalizers $U'P$ is in fact a sheaf. Temporarily considering $U'\colon \underline{P}(X,\underline{B}) \to \underline{P}(X,\underline{A})$ and letting $F'\colon \underline{P}(X,\underline{A}) \to \underline{P}(X,\underline{B})$ be its left adjoint (adjoints lift to functor categories), we easily see that the left adjoint of $U'\colon \underline{F}(X,\underline{B}) \to \underline{F}(X,\underline{A})$ is the composition $\underline{F}(X,\underline{A}) \xrightarrow{i} \underline{P}(X,\underline{A}) \xrightarrow{F'} \underline{P}(X,\underline{B}) \xrightarrow{\check{\Psi}\Psi} \underline{F}(X,\underline{B})$. Relabelling, we have an adjoint pair $F' \dashv U'\colon \underline{F}(X,\underline{B}) \to \underline{F}(X,\underline{A})$. Next notice that since U commutes with directed colimits, the diagram:

$$\begin{array}{ccc} \underline{F}(X,\underline{B}) & \xrightarrow{S'i} & \underline{B}^{|X|} \\ {\scriptstyle U'}\downarrow & & \downarrow{\scriptstyle U^{|X|}} \\ \underline{F}(X,\underline{A}) & \xrightarrow{Si} & \underline{A}^{|X|} \end{array}$$

commutes. We now verify the condition of Beck's Precise Tripleableness Theorem for U' [3], [6].

i) U' reflects isomorphisms, for let $f\colon P' \to P$ be a map in $\underline{F}(X,\underline{B})$ such that $U'f$ is an isomorphism. Then $U^{|X|}\, S'if \approx SU'f$ is an isomorphism, and since $U^{|X|}$ is tripleable, $S'if$ is an isomorphism. Thus f is an isomorphism, because $S'i$ is cotripleable.

ii) U' preserves coequalizers of U'-split pair say $U'P' \rightleftarrows U'P \rightleftarrows C$, and $P' \rightrightarrows P \longrightarrow D$ the coequalizer in $\underline{F}(X,\underline{B})$. We must show that $U'D \approx C$. Now

$S'iP' \rightrightarrows S'iP$ is a $U^{|X|}$ -split pair, namely $U^{|X|}S'iP' \rightrightarrows U^{|X|}S'iP \rightleftarrows SiC$.
But $S'i$ preserves coequalizers, so that $S'iP' \rightrightarrows S'iP \longrightarrow S'D$ is a coequalizer. It is preserved by the tripleable $U^{|X|}$, and hence $U^{|X|}S'iD \approx SiC$. But $U^{|X|}S'iD \approx SiU'D$, so $SiC \approx SiU'D$. Since Si is cotripleable it reflects isomorphisms, and therefore $C \approx U'D$. It follows that U' is tripleable.

IV. Interpretation and Examples

The situation is that of Theorem III.6. For notational convenience we write $U = U^{|X|}$, $F = F^{|X|}$, and $S = Si$. Thus we have a triple $\mathbb{T} = (UF,\eta,\mu)$ and a cotriple $\mathbb{G} = (SQ,\varepsilon,\delta)$ defined on $\underline{A}^{|X|}$. For each $\{A_x\}$ in $\underline{A}^{|X|}$ and each open subset V of X we define ϕ_V in the diagram:

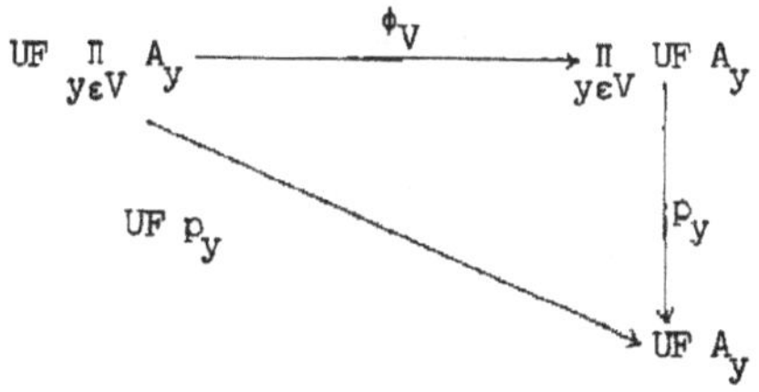

Using these ϕ's , we define $\lambda_x\{A_y\}$ to be the composition:

$$UF \operatorname*{colim}_{x\in V} \prod_{y\in V} A_y \xrightarrow{\ \cong\ } \operatorname*{colim}_{x\in V} UF \prod_{y\in V} A_y \xrightarrow{\ \operatorname{colim} \phi_V\ } \operatorname*{colim}_{x\in V} \prod_{y\in V} UF\, A_y ,$$

thus getting a natural transformation λ: $UFSQ \rightarrow SQUF$. This λ is a mixed distributive law from $\mathbb{T}$ to $\mathbb{G}$, and we digress briefly to talk about such laws.

A mixed distributive law from triple $\mathbb{T}$ to cotriple $\mathbb{G}$ [2] is a natural transformation λ: $TG \rightarrow GT$ satisfying:

i) $G\lambda \cdot \lambda G \cdot T\delta = \delta T \cdot \lambda$

ii) $G\mu \cdot \lambda T \cdot T\lambda = \lambda \cdot \mu G$

iii) $\lambda \cdot \eta G = G\eta$

iv) $\varepsilon T \cdot \lambda = T\varepsilon$.

Theorem IV.1. Let $\mathbb{T}$ and $\mathbb{G}$ be defined on $\underline{A}$. Then the following are equivalent:

i) There exists a mixed distributive law $\lambda\colon\ TG \to GT$.

ii) There exists a cotriple $\mathbb{G}'$ on $\underline{A}^{\mathbb{T}}$ such that
$U^{\mathbb{T}}G' = GU^{\mathbb{T}}$, $U^{\mathbb{T}}\varepsilon' = \varepsilon U^{\mathbb{T}}$, $U^{\mathbb{T}}\delta' = \delta U^{\mathbb{T}}$.

iii) There exists a triple $\mathbb{T}'$ on $\underline{A}_{\mathbb{G}}$ such that
$S_{\mathbb{G}}T' = TS_{\mathbb{G}}$, $S_{\mathbb{G}}\eta' = \eta S_{\mathbb{G}}$, $S_{\mathbb{G}}\mu' = \mu S_{\mathbb{G}}$.

The correspondences $\lambda \leftrightarrow \mathbb{G}' \leftrightarrow \mathbb{T}'$ are one-to-one.

Proof: Given a mixed distributive law $\lambda\colon\ TG \to GT$, define $G'(A,\xi) = (GA, G\xi\cdot\lambda A)$, $\varepsilon' = \varepsilon$, $\delta' = \delta$. Given $\mathbb{G}'$ on $\underline{A}^{\mathbb{T}}$ define $\lambda\colon\ TG \to GT$ to be the composition $TG \xrightarrow{TG\eta} TGT = U^{\mathbb{T}}F^{\mathbb{T}}U^{\mathbb{T}}G'F^{\mathbb{T}} \xrightarrow{U^{\mathbb{T}}\theta G'F^{\mathbb{T}}} U^{\mathbb{T}}G'F^{\mathbb{T}}$ where $\theta\colon\ F^{\mathbb{T}}U^{\mathbb{T}} \to \underline{A}^{\mathbb{T}}$ is the adjunction morphism. Thus i and ii are equivalent. Moreover, it is easy to see that these passages are mutually inverse. A similar proof shows the equivalence of i and iii .

It follows from this theorem that if we are given a mixed distributive law $\lambda\colon\ TG \to GT$ then $(\underline{A}^{\mathbb{T}})_{\mathbb{G}'} = (\underline{A}_{\mathbb{G}})^{\mathbb{T}'}$. In fact, an object in $(\underline{A}^{\mathbb{T}})_{\mathbb{G}'}$ is a three-tuple $((A,\xi),\alpha)$ where:

i) $\xi\cdot\mu A = \xi\cdot T\xi$

ii) $\xi\cdot\eta A = A$

iii) $G\alpha\cdot\alpha = \delta A\cdot\alpha$

iv) $\varepsilon A\cdot\alpha = A$

v) $G\xi\cdot\lambda A\cdot T\alpha = \alpha\cdot\xi$.

A morphism $f\colon\ ((A,\xi),\alpha) \to ((A',\xi'),\alpha')$ is an $\underline{A}$-morphism $f\colon\ A \to A'$ such that:

i) $\xi'\cdot Tf = f\cdot\xi$

ii) $\alpha'\cdot f = Gf\cdot\alpha$.

The conditions for objects and maps in $(\underline{A}_{\mathbb{G}})^{\mathbb{T}'}$ are the same, and we denote this category by $\underline{A}_{\mathbb{G}}^{\mathbb{T}}$, λ being understood. Equivalently, $\underline{A}_{\mathbb{G}}^{\mathbb{T}}$ is the full subcategory of the pullback of $S_{\mathbb{G}}\colon\ \underline{A}_{\mathbb{G}} \to \underline{A}$ and $U^{\mathbb{T}}\colon\ \underline{A}^{\mathbb{T}} \to \underline{A}$ consisting of those (A,ξ,α) such that $G\xi\cdot\lambda A\cdot T\alpha = \alpha\cdot\xi$.

We now return to our mixed distributive law λ: UFSQ $\to$ SQUF for sheaves. From the foregoing discussion it follows that $\underline{F}(X,\underline{B}) \approx (\underline{A}^{|X|})^{\mathbb{T}}_{\mathbb{G}}$. We want to interpret this equivalence in the case $\underline{A}$ = $\underline{\text{Sets}}$. By Theorem III.3, a sheaf of sets is equivalent to a coalgebra $\{(\{A_x\},\{\alpha_x\})|x \text{ in } X\}$. This means that we have one "stalk" A_x for each x in X , together with instructions (the α_x's) on how to fit them together. Specifically, if we let $+A_x$ be the disjoint union of the A_x , and $[V,+A_x]$ the set of all functions f from V to $+A_x$ such that f(v) is in A_v , then $SQ\{A_x\} = \{\text{colim}[V,+A_x]|$the colimit is taken over all open sets V which contain x$\}$. Thus if a_x is in A_x then $\alpha_x(a_x)$ is represented by a function $\alpha_x(a_x)$ in $[V,+A_x]$ for some open neighborhood V of x . The two conditions that $\{\alpha_x\}$ be a costructure are then:

i) $[\alpha_x(a_x)](x) = a_x$, and

ii) $\alpha_x(a_x) = \alpha_v([\alpha_x(a_x)](v))$ for all v in some sufficiently small neighborhood V of x .

Thus if v is close to x in X then A_x and A_v depend on each other, and the α_x's tell us how. This makes precise what sheaf theorists mean when they say that the stalks vary continuously over X . In fact, if we let $U(a_x) = \{\alpha_x(a_x)(V)|V$ is a neighborhood of x such that ii above holds for all v in V$\}$ then $U(a_x)$ is a local base at a_x , and the topology on $+A_x$ which these local bases induce is precisely the one which sheaf theorists talk about.

A sheaf with values in $\underline{\text{Sets}}^{\mathbb{T}}$($\mathbb{T}$ finitary) can now be easily described in the new context. It will be a sheaf $(\{A_x\},\{\alpha_x\})$ of sets (in the above sense) each stalk (A_x,ξ_x) of which is a $\mathbb{T}$-algebra, and such that the sheaf and algebra structures are compatible. This compatibility condition is requirement (v) above. It says that with respect to the topology we have defined, the $\mathbb{T}$-algebra operations are continuous; or equivalently, that the sheaf costructure mappings are $\mathbb{T}$-algebra homomorphisms.

We conclude by offering examples of how this new description can be used to construct some well-known sheaves.

Example 1. The constant sheaf of integers. For each x in X let $A_x = Z$ = the integers. Define α_x: $A_x \to \text{colim}[V,+A_x]$ to be such that $\alpha_x(z)$ is represented by the function f: $X \to +A_x$ with $f(x) = z$ for all x in X.

Example 2. The sheaf of germs of holomorphic functions. Let X be an open subset of the complex plane and let A_x be the set of all power series f_x such that there is a neighborhood V' of x on which f_x converges. Let α_x: $A_x \to \text{colim}$ $[V,+A_x]$ be defined by the condition that $\alpha_x(f_x)$ is represented by the function g: $V' \to +A_x$ such that $g(v)$ = the power series expansion of f_x around the point v in V'.

Example 3. The affine scheme of a ring. Let R be a commutative ring and $X = \text{Spec}(R)$ with the Zariski topology. Recall that if, for each $r \varepsilon R$, we let $D(r) = \{x \varepsilon X | r \notin x\}$ and $U(x) = \{D(r) | x \varepsilon D(r)\}$ then the $U(x)$ form a local basis for the Zariski topology in X. For each x in X, let A_x be the local ring of R with respect to the prime ideal x. Note that colim $[V,+A_x]$ is isomorphic to $\text{colim}[D(r),+A_x]$ where the colimit is taken over all $D(r)$ in $U(x)$; this is because $U(x)$ is cofinal in the set of all open subsets of X which contain x (since $U(x)$ is a local base at x). Thus it suffices to define α_x: $A_x \to \text{colim}$ $[D(r),+A_x]$ where the colimit is taken over $U(x)$. For $\frac{t}{s}$ in A_x we define $\alpha_x(\frac{t}{s})$ to be represented by the function g: $D(s) \to +A_x$ where $g(y) = \frac{t}{s}$.

Example 4. Sheaf of germs of homomorphisms. Notice that if $\{(A_x,\alpha_x)\}$ is a sheaf of sets on X and Y is any subspace of X then $\{(A_x,\alpha_x)|x \varepsilon Y\}$ is a sheaf of sets on Y (called the restriction of $\{(A_x,\alpha_x)\}$ to Y.) Given two sheaves of sets $\{(A_x,\alpha_x)\}$ and $\{(A_x{}',\alpha_x{}')\}$ on X, we let $C_x = \{f\colon A_x \to \text{colim}[V,+A_x] |$ there is an open neighborhood V_f of x and a sheaf morphism $\{h_x\}$: $\{(A_x,\alpha_x)|x\varepsilon V_f\} \to \{(A_x{}',\alpha_x{}')|x\varepsilon V_f\}$ such that $\alpha_x{}' \cdot h_x = f$. Define γ_x: $C_x \to \text{colim}[V,+C_x]$ to be that function such that $\gamma_x(f)$ is represented by the function g: $V_f \to +C_x$, where $g(v) = \alpha_v{}' \cdot h_v$ for each v in V_f. The sheaf $\{(C_x,\gamma_x)\}$ is called the sheaf of germs of homomorphisms from $\{(A_x,\alpha_x)\}$ to $\{(A_x{}',\alpha_x{}')\}$.

References

[1] M. Barr, Non-abelian Full Embedding, this volume.

[2] J. Beck, Distributive Laws, Springer-Verlag Lecture Notes in Mathematics, Volume 80 (1969), pp. 119-140.

[3] J. Beck, The Tripleableness Theorem, unpublished manuscript, Cornell University, 1967.

[4] R. Godement, Topologie Algébrique et Théorie des Faisceaux, Hermann, Paris (1964).

[5] P.-A. Grillet, Inductive Limits and Categories with Decompositions, this volume.

[6] E. Manes, A Triple-Theoretic Constuction of Compact Algebras, Springer-Verlag Lecture Notes in Mathematics, Volume 80 (1969), pp. 91-118.

[7] B. Mitchell, Theory of Categories, Academic Press, New York (1965).

Lecture Notes in Mathematics

Comprehensive leaflet on request

Vol. 38: R. Berger, R. Kiehl, E. Kunz und H.-J. Nastold, Differentialrechnung in der analytischen Geometrie IV, 134 Seiten. 1967 DM 12,–

Vol. 39: Séminaire de Probabilités I. II, 189 pages. 1967. DM 14,–

Vol. 40: J. Tits, Tabellen zu den einfachen Lie Gruppen und ihren Darstellungen. VI, 53 Seiten. 1967. DM 6.80

Vol. 41: A. Grothendieck, Local Cohomology. VI, 106 pages. 1967. DM 10,–

Vol. 42: J. F. Berglund and K. H. Hofmann, Compact Semitopological Semigroups and Weakly Almost Periodic Functions. VI, 160 pages. 1967. DM 12,–

Vol. 43: D. G. Quillen, Homotopical Algebra. VI, 157 pages. 1967. DM 14,–

Vol. 44: K. Urbanik, Lectures on Prediction Theory. IV, 50 pages. 1967. DM 5,80

Vol. 45: A. Wilansky, Topics in Functional Analysis. VI, 102 pages. 1967. DM 9,60

Vol. 46: P. E. Conner, Seminar on Periodic Maps. IV, 116 pages. 1967. DM 10,60

Vol. 47: Reports of the Midwest Category Seminar I. IV, 181 pages. 1967. DM 14,80

Vol. 48: G. de Rham, S. Maumary et M. A. Kervaire, Torsion et Type Simple d'Homotopie. IV, 101 pages. 1967. DM 9,60

Vol. 49: C. Faith, Lectures on Injective Modules and Quotient Rings. XVI, 140 pages. 1967. DM 12,80

Vol. 50: L. Zalcman, Analytic Capacity and Rational Approximation. VI, 155 pages. 1968. DM 13.20

Vol. 51: Séminaire de Probabilités II. IV, 199 pages. 1968. DM 14,–

Vol. 52: D. J. Simms, Lie Groups and Quantum Mechanics. IV, 90 pages. 1968. DM 8,–

Vol. 53: J. Cerf, Sur les difféomorphismes de la sphère de dimension trois (Γ_4 = O). XII, 133 pages. 1968. DM 12,–

Vol. 54: G. Shimura, Automorphic Functions and Number Theory. VI, 69 pages. 1968. DM 8,–

Vol. 55: D. Gromoll, W. Klingenberg und W. Meyer, Riemannsche Geometrie im Großen. VI, 287 Seiten. 1968. DM 20,–

Vol. 56: K. Floret und J. Wloka, Einführung in die Theorie der lokalkonvexen Räume. VIII, 194 Seiten. 1968. DM 16,–

Vol. 57: F. Hirzebruch und K. H. Mayer, O (n)-Mannigfaltigkeiten, exotische Sphären und Singularitäten. IV, 132 Seiten. 1968. DM 10,80

Vol. 58: Kuramochi Boundaries of Riemann Surfaces. IV, 102 pages. 1968. DM 9,60

Vol. 59: K. Jänich, Differenzierbare G-Mannigfaltigkeiten. VI, 89 Seiten. 1968. DM 8,–

Vol. 60: Seminar on Differential Equations and Dynamical Systems. Edited by G. S. Jones. VI, 106 pages. 1968. DM 9,60

Vol. 61: Reports of the Midwest Category Seminar II. IV, 91 pages. 1968. DM 9,60

Vol. 62: Harish-Chandra, Automorphic Forms on Semisimple Lie Groups X, 138 pages. 1968. DM 14,–

Vol. 63: F. Albrecht, Topics in Control Theory. IV, 65 pages. 1968. DM 6,80

Vol. 64: H. Berens, Interpolationsmethoden zur Behandlung von Approximationsprozessen auf Banachräumen. VI, 90 Seiten. 1968. DM 8,–

Vol. 65: D. Kölzow, Differentiation von Maßen. XII, 102 Seiten. 1968. DM 8,–

Vol. 66: D. Ferus, Totale Absolutkrümmung in Differentialgeometrie und -topologie. VI, 85 Seiten. 1968. DM 8,–

Vol. 67: F. Kamber and P. Tondeur, Flat Manifolds. IV, 53 pages. 1968. DM 5,80

Vol. 68: N. Boboc et P. Mustată, Espaces harmoniques associés aux opérateurs différentiels linéaires du second ordre de type elliptique. VI, 95 pages. 1968. DM 8,60

Vol. 69: Seminar über Potentialtheorie. Herausgegeben von H. Bauer. VI, 180 Seiten. 1968. DM 14,80

Vol. 70: Proceedings of the Summer School in Logic. Edited by M. H. Löb. IV, 331 pages. 1968. DM 20,–

Vol. 71: Séminaire Pierre Lelong (Analyse), Année 1967 – 1968. VI, 190 pages. 1968. DM 14,

Vol. 72: The Syntax and Semantics of Infinitary Languages. Edited by J. Barwise. IV, 268 pages. 1968. DM 18,–

Vol. 73: P. E. Conner, Lectures on the Action of a Finite Group. IV, 123 pages. 1968. DM 10,–

Vol. 74: A. Fröhlich, Formal Groups. IV, 140 pages. 1968. DM 12,–

Vol. 75: G. Lumer, Algèbres de fonctions et espaces de Hardy. VI, 80 pages. 1968. DM 8,–

Vol. 76: R. G. Swan, Algebraic K-Theory. IV, 262 pages. 1968. DM 18,–

Vol. 77: P.-A. Meyer, Processus de Markov: la frontière de Martin. IV, 123 pages. 1968. DM 10,–

Vol. 78: H. Herrlich, Topologische Reflexionen und Coreflexionen. XVI, 166 Seiten. 1968. DM 12,–

Vol. 79: A. Grothendieck, Catégories Cofibrées Additives et Complexe Cotangent Relatif. IV, 167 pages. 1968. DM 12,–

Vol. 80: Seminar on Triples and Categorical Homology Theory. Edited by B. Eckmann. IV, 398 pages. 1969. DM 20,–

Vol. 81: J.-P. Eckmann et M. Guenin, Méthodes Algébriques en Mécanique Statistique. VI, 131 pages. 1969. DM 12,–

Vol. 82: J. Wloka, Grundräume und verallgemeinerte Funktionen. VIII, 131 Seiten. 1969. DM 12,–

Vol. 83: O. Zariski, An Introduction to the Theory of Algebraic Surfaces. IV, 100 pages. 1969. DM 8,–

Vol. 84: H. Lüneburg, Transitive Erweiterungen endlicher Permutationsgruppen. IV, 119 Seiten. 1969. DM 10.–

Vol. 85: P. Cartier et D. Foata, Problèmes combinatoires de commutation et réarrangements. IV, 88 pages. 1969. DM 8,–

Vol. 86: Category Theory, Homology Theory and their Applications I. Edited by P. Hilton. VI, 216 pages. 1969. DM 16,–

Vol. 87: M. Tierney, Categorical Constructions in Stable Homotopy Theory. IV, 65 pages. 1969. DM 6,–

Vol. 88: Séminaire de Probabilités III. IV, 229 pages. 1969. DM 18,–

Vol. 89: Probability and Information Theory. Edited by M. Behara, K. Krickeberg and J. Wolfowitz. IV, 256 pages. 1969. DM 18,–

Vol. 90: N. P. Bhatia and O. Hajek, Local Semi-Dynamical Systems. II, 157 pages. 1969. DM 14,–

Vol. 91: N. N. Janenko, Die Zwischenschrittmethode zur Lösung mehrdimensionaler Probleme der mathematischen Physik. VIII, 194 Seiten. 1969. DM 16,80

Vol. 92: Category Theory, Homology Theory and their Applications II. Edited by P. Hilton. V, 308 pages. 1969. DM 20,–

Vol. 93: K. R. Parthasarathy, Multipliers on Locally Compact Groups. III, 54 pages. 1969. DM 5,60

Vol. 94: M. Machover and J. Hirschfeld, Lectures on Non-Standard Analysis. VI, 79 pages. 1969. DM 6,–

Vol. 95: A. S. Troelstra, Principles of Intuitionism. II, 111 pages. 1969. DM 10,–

Vol. 96: H.-B. Brinkmann und D. Puppe, Abelsche und exakte Kategorien, Korrespondenzen. V, 141 Seiten. 1969. DM 10,–

Vol. 97: S. O. Chase and M. E. Sweedler, Hopf Algebras and Galois theory. II, 133 pages. 1969. DM 10,–

Vol. 98: M. Heins, Hardy Classes on Riemann Surfaces. III, 106 pages. 1969. DM 10,–

Vol. 99: Category Theory, Homology Theory and their Applications III. Edited by P. Hilton. IV, 489 pages. 1969. DM 24,–

Vol. 100: M. Artin and B. Mazur, Etale Homotopy. II, 196 Seiten. 1969. DM 12,–

Vol. 101: G. P. Szegö et G. Treccani, Semigruppi di Trasformazioni Multivoche. VI, 177 pages. 1969. DM 14,–

Vol. 102: F. Stummel, Rand- und Eigenwertaufgaben in Sobolewschen Räumen. VIII, 386 Seiten. 1969. DM 20,–

Vol. 103: Lectures in Modern Analysis and Applications I. Edited by C. T. Taam. VII, 162 pages. 1969. DM 12,–

Vol. 104: G. H. Pimbley, Jr., Eigenfunction Branches of Nonlinear Operators and their Bifurcations. II, 128 pages. 1969. DM 10,–

Vol. 105: R. Larsen, The Multiplier Problem. VII, 284 pages. 1969. DM 18,–

Vol. 106: Reports of the Midwest Category Seminar III. Edited by S. Mac Lane. III, 247 pages. 1969. DM 16,–

Vol. 107: A. Peyerimhoff, Lectures on Summability. III, 111 pages. 1969. DM 8,–

Vol. 108: Algebraic K-Theory and its Geometric Applications. Edited by R. M.F. Moss and C. B. Thomas. IV, 86 pages. 1969. DM 6,–

Vol. 109: Conference on the Numerical Solution of Differential Equations. Edited by J. Ll. Morris. VI, 275 pages. 1969. DM 18,–

Vol. 110: The Many Facets of Graph Theory. Edited by G. Chartrand and S. F. Kapoor. VIII, 290 pages. 1969. DM 18,–

Vol. 111: K. H. Mayer, Relationen zwischen charakteristischen Zahlen. III, 99 Seiten. 1969. DM 8,–

Vol. 112: Colloquium on Methods of Optimization. Edited by N. N. Moiseev. IV, 293 pages. 1970. DM 18,–

Vol. 113: R. Wille, Kongruenzklassengeometrien. III, 99 Seiten. 1970 DM 8,–

Vol. 114: H. Jacquet and R. P. Langlands, Automorphic Forms on GL (2). VII, 548 pages. 1970. DM 24,–

Vol. 115: K. H. Roggenkamp and V. Huber-Dyson, Lattices over Orders I. XIX, 290 pages. 1970. DM 18,–

Vol. 116: Séminaire Pierre Lelong (Analyse) Année 1969. IV, 195 pages. 1970. DM 14,–

Vol. 117: Y. Meyer, Nombres de Pisot, Nombres de Salem et Analyse Harmonique. 63 pages. 1970. DM 6.–

Vol. 118: Proceedings of the 15th Scandinavian Congress, Oslo 1968. Edited by K. E. Aubert and W. Ljunggren. IV, 162 pages. 1970. DM 12,–

Vol. 119: M. Raynaud, Faisceaux amples sur les schémas en groupes et les espaces homogènes. III, 219 pages. 1970. DM 14,–

Vol. 120: D. Siefkes, Büchi's Monadic Second Order Successor Arithmetic. XII, 130 Seiten. 1970. DM 12,–

Vol. 121: H. S. Bear, Lectures on Gleason Parts. III, 47 pages. 1970. DM 6,–

Vol. 122: H. Zieschang, E. Vogt und H.-D. Coldewey, Flächen und ebene diskontinuierliche Gruppen. VIII, 203 Seiten. 1970. DM 16,–

Vol. 123: A. V. Jategaonkar, Left Principal Ideal Rings. VI, 145 pages. 1970. DM 12,–

Vol. 124: Séminare de Probabilités IV. Edited by P. A. Meyer. IV, 282 pages. 1970. DM 20,–

Vol. 125: Symposium on Automatic Demonstration. V, 310 pages. 1970. DM 20,–

Vol. 126: P. Schapira, Théorie des Hyperfonctions. XI, 157 pages. 1970. DM 14,–

Vol. 127: I. Stewart, Lie Algebras. IV, 97 pages. 1970. DM 10,–

Vol. 128: M. Takesaki, Tomita's Theory of Modular Hilbert Algebras and its Applications. II, 123 pages. 1970. DM 10,–

Vol. 129: K. H. Hofmann, The Duality of Compact Semigroups and C*-Bigebras. XII, 142 pages. 1970. DM 14,–

Vol. 130: F. Lorenz, Quadratische Formen über Körpern. II, 77 Seiten. 1970. DM 8,–

Vol. 131: A Borel et al., Seminar on Algebraic Groups and Related Finite Groups. VII, 321 pages. 1970. DM 22,–

Vol. 132: Symposium on Optimization. III, 348 pages. 1970. DM 22,–

Vol. 133: F. Topsøe, Topology and Measure. XIV, 79 pages. 1970. DM 8,–

Vol. 134: L. Smith, Lectures on the Eilenberg-Moore Spectral Sequence. VII, 142 pages. 1970. DM 14,–

Vol. 135: W. Stoll, Value Distribution of Holomorphic Maps into Compact Complex Manifolds. II, 267 pages. 1970. DM 18,–

Vol. 136: M. Karoubi et al., Séminaire Heidelberg-Saarbrücken-Strasbourg sur la K-Théorie. IV, 264 pages. 1970. DM 18,–

Vol. 137: Reports of the Midwest Category Seminar IV. Edited by S. MacLane. III, 139 pages. 1970. DM 12,–

Vol. 138: D. Foata et M. Schützenberger, Théorie Géométrique des Polynômes Eulériens. V, 94 pages. 1970. DM 10,–

Vol. 139: A. Badrikian, Séminaire sur les Fonctions Aléatoires Linéaires et les Mesures Cylindriques. VII, 221 pages. 1970. DM 18,–

Vol. 140: Lectures in Modern Analysis and Applications II. Edited by C. T. Taam. VI, 119 pages. 1970. DM 10,–

Vol. 141: G. Jameson, Ordered Linear Spaces. XV, 194 pages. 1970. DM 16,–

Vol. 142: K. W. Roggenkamp, Lattices over Orders II. V, 388 pages. 1970. DM 22,–

Vol. 143: K. W. Gruenberg, Cohomological Topics in Group Theory. XIV, 275 pages. 1970. DM 20,–

Vol. 144: Seminar on Differential Equations and Dynamical Systems, II. Edited by J. A. Yorke. VIII, 268 pages. 1970. DM 20,–

Vol. 145: E. J. Dubuc, Kan Extensions in Enriched Category Theory. XVI, 173 pages. 1970. DM 16,–

Vol. 146: A. B. Altman and S. Kleiman, Introduction to Grothendieck Duality Theory. II, 192 pages. 1970. DM 18,–

Vol. 147: D. E. Dobbs, Cech Cohomological Dimensions for Commutative Rings. VI, 176 pages. 1970. DM 16,–

Vol. 148: R. Azencott, Espaces de Poisson des Groupes Localement Compacts. IX, 141 pages. 1970. DM 14,–

Vol. 149: R. G. Swan and E. G. Evans, K-Theory of Finite Groups and Orders. IV, 237 pages. 1970. DM 20,–

Vol. 150: Heyer, Dualität lokalkompakter Gruppen. XIII, 372 Seiten. 1970. DM 20,–

Vol. 151: M. Demazure et A. Grothendieck, Schémas en Groupes I. (SGA 3). XV, 562 pages. 1970. DM 24,–

Vol. 152: M. Demazure et A. Grothendieck, Schémas en Groupes II. (SGA 3). IX, 654 pages. 1970. DM 24,–

Vol. 153: M. Demazure et A. Grothendieck, Schémas en Groupes III. (SGA 3). VIII, 529 pages. 1970. DM 24,–

Vol. 154: A. Lascoux et M. Berger, Variétés Kähleriennes Compactes. VII, 83 pages. 1970. DM 8,–

Vol. 155: Several Complex Variables I, Maryland 1970. Edited by J. Horváth. IV, 214 pages. 1970. DM 18,–

Vol. 156: R. Hartshorne, Ample Subvarieties of Algebraic Varieties. XIV, 256 pages. 1970. DM 20,–

Vol. 157: T. tom Dieck, K. H. Kamps und D. Puppe, Homotopietheorie. VI, 265 Seiten. 1970. DM 20,–

Vol. 158: T. G. Ostrom, Finite Translation Planes. IV. 112 pages. 1970. DM 10,–

Vol. 159: R. Ansorge und R. Hass. Konvergenz von Differenzenverfahren für lineare und nichtlineare Anfangswertaufgaben. VIII, 145 Seiten. 1970. DM 14,–

Vol. 160: L. Sucheston, Constributions to Ergodic Theory and Probability. VII, 277 pages. 1970. DM 20,–

Vol. 161: J. Stasheff, H-Spaces from a Homotopy Point of View. VI, 95 pages. 1970. DM 10,–

Vol. 162: Harish-Chandra and van Dijk, Harmonic Analysis on Reductive p-adic Groups. IV, 125 pages. 1970. DM 12,–

Vol. 163: P. Deligne, Equations Différentielles à Points Singuliers Reguliers. III, 133 pages. 1970. DM 12,–

Vol. 164: J. P. Ferrier, Seminaire sur les Algebres Complètes. II, 69 pages. 1970. DM 8,–

Vol. 165: J. M. Cohen, Stable Homotopy. V, 194 pages. 1970. DM 16,–

Vol. 166: A. J. Silberger, PGL_2 over the p-adics: its Representations, Spherical Functions, and Fourier Analysis. VII, 202 pages. 1970. DM 18,–

Vol. 167: Lavrentiev, Romanov and Vasiliev, Multidimensional Inverse Problems for Differential Equations. V, 59 pages. 1970. DM 10,–

Vol. 168: F. P. Peterson, The Steenrod Algebra and its Applications: A conference to Celebrate N. E. Steenrod's Sixtieth Birthday. VII, 317 pages. 1970. DM 22,–

Vol. 169: M. Raynaud, Anneaux Locaux Henséliens. V, 129 pages. 1970. DM 12,–

Vol. 170: Lectures in Modern Analysis and Applications III. Edited by C. T. Taam. VI, 213 pages. 1970. DM 18,–

Vol. 171: Set-Valued Mappings, Selections and Topological Properties of 2^X. Edited by W. M. Fleischman. X, 110 pages. 1970. DM 12,–

Vol. 172: Y.-T. Siu and G. Trautmann, Gap-Sheaves and Extension of Coherent Analytic Subsheaves. V, 172 pages. 1971. DM 16,–

Vol. 173: J. N. Mordeson and B. Vinograde, Structure of Arbitrary Purely Inseparable Extension Fields. IV, 138 pages. 1970. DM 14,–

Vol. 174: B. Iversen, Linear Determinants with Applications to the Picard Scheme of a Family of Algebraic Curves. VI, 69 pages. 1970. DM 8,–

Vol. 175: M. Brelot, On Topologies and Boundaries in Potential Theory. VI, 176 pages. 1971. DM 18,–

Vol. 176: H. Popp, Fundamentalgruppen algebraischer Mannigfaltigkeiten. IV, 154 Seiten. 1970. DM 16,–

Vol. 177: J. Lambek, Torsion Theories, Additive Semantics and Rings of Quotients. VI, 94 pages. 1971. DM 12,–

Vol. 178: Th. Bröcker und T. tom Dieck, Kobordismentheorie. XVI, 191 Seiten. 1970. DM 18,–

Vol. 179: Seminaire Bourbaki – vol. 1968/69. Exposés 347-363. IV. 295 pages. 1971. DM 22,–

Vol. 180: Séminaire Bourbaki – vol. 1969/70. Exposés 364-381. IV, 310 pages. 1971. DM 22,–

Vol. 181: F. DeMeyer and E. Ingraham, Separable Algebras over Commutative Rings. V, 157 pages. 1971. DM 16.–